Upskilling for the Age of AI Automation

Edited by

Rishikaysh Kaakandikar
Management Sciences, SaiBalaji International Institute of Management Sciences Pune, India

Keshav Kaushik
Amity School of Engineering and Technology Amity University Mohali, Punjab India

Supriya Pathak
Faculty of Management and Commerce Oriental University, Indore Madhya Pradesh, India

&

Madhavi Shamkuwar
Zeal Institute of Business Administration Computer Application & Research Pune, India

Upskilling for the Age of AI Automation

Editors: Rishikaysh Kaakandikar, Keshav Kaushik, Supriya Pathak and

Madhavi Shamkuwar

ISBN (Online): 979-8-89881-261-4

ISBN (Print): 979-8-89881-262-1

ISBN (Paperback): 979-8-89881-263-8

Published by Bentham Science Publishers Pte. Ltd. Singapore,

in collaboration with Eureka Conferences, USA.

First published in 2026.

Bentham Science Publishers Pte. Ltd.
No. 9 Raffles Place
Office No. 26-01
Singapore 048619
Singapore
Email: subscriptions@benthamscience.net

CONTENTS

FOREWORD

The twenty-first century has ushered in a new era, defined by Artificial Intelligence, automation, and rapid digital transformation. As industries reshape, work evolves, and skills become obsolete faster than ever before, the need to upskill is no longer optional—it is existential. "Upskilling for the Age of AI and Automation" captures this imperative with depth, clarity, and vision.

From the opening chapter, which lays a strong foundation for continuous learning, to the later discussions on inclusivity, ethics, and generational adaptation, this book provides a panoramic view of what it takes to thrive in a world where machines learn and adapt alongside us. The early chapters map the essential skillsets for the next decade—spanning technical literacy, data fluency, creativity, emotional intelligence, and sustainability awareness—while grounding these in real-world contexts across industries.

As McKinsey & Company's recent Future of Work reports remind us, as many as 375 million workers globally—or about 14% of the world's workforce—may need to switch occupational categories by 2030 due to automation. Moreover, nine out of ten executives now say they face skill gaps or expect them to emerge within five years. Yet only a small fraction—less than 40% of organizations—believe they are effectively addressing this challenge. This widening skills gap underlines the urgency of initiatives that combine human adaptability with continuous technological learning—the very ethos of this book.

The middle chapters make a compelling case for balancing hard and soft skills, simplifying complex technical ideas like machine learning and cloud computing for wider audiences. They show that while data and algorithms may power tomorrow's economy, empathy, leadership, and communication will define its human core. Later sections skill-fully weave in generational and gender perspectives, reminding readers that reskilling must be inclusive—embracing digital immigrants as well as digital natives, men and women alike.

What sets this book apart is its multidisciplinary scope. It integrates management, education, psychology, and technology into a coherent narrative on how individuals, institutions, and societies can prepare for a future shaped by AI. Each chapter combines academic rigour with practical insight—making the text equally valuable for students, educators, HR professionals, policymakers, and business leaders.

In a time when learning itself is being redefined, "Upskilling for the Age of AI and Automation" serves as both a guidebook and a call to action. It reminds us that while machines may augment our work, the uniquely human capacity to learn, adapt, and collaborate will always remain our greatest asset.

Ganesh M. Kakandikar
Department of Innovation, Startup and Collaboration
Dr. Vishwanath Karad MIT World Peace University
Pune, India

PREFACE

Artificial Intelligence (AI) has transformed from a hypothetical idea to become an essential part of the modern transformative age. Being a disruptive technology and an enabling technology simultaneously, AI is transforming the borders of various areas, including agriculture, healthcare, education, and so forth. The need to upskill the global workforce becomes very critical in this technological transition. The provision of employees with the necessary skills is not just essential to face the challenges of AI-based automation, but also to take advantage of all its potential in relation to the development of innovations, the deepening of productivity, and the growth of an inclusive society.

This book is designed as an all-purpose guide to individuals, teachers, and institutions. It aims at offering practical insights, an industry-based model of skill development, and future-oriented plans to develop professionals who are adequately equipped to operate with the latest technological paradigms.

The editors would like to express their inmost appreciation to all contributors for their intellectual input, and to Bentham Science Publishers, who have continued to support this work without any hesitation.

The book challenges the readers to go through the critical adventure of AI-powered upskilling, a call that is considered essential to succeed in the fast-changing technologized environment and remain competitive in all professional spheres.

Rishikaysh Kaakandikar
Management Sciences, SaiBalaji
International Institute of Management Sciences
Pune, India

Keshav Kaushik
Amity School of Engineering and Technology
Amity University Mohali, Punjab
India

Supriya Pathak
Faculty of Management and Commerce
Oriental University, Indore
Madhya Pradesh, India

&

Madhavi Shamkuwar
Zeal Institute of Business Administration
Computer Application & Research
Pune, India

List of Contributors

Ajinkya Sanjay Bhalerao	Sinhgad Institute of Management and Computer Application, Pune, India
Anju Rohilla	Department of Management Studies, SRM University, Delhi-NCR, Sonepat, India
Babasaheb Jadhav	Global Business School & Research Centre, Dr. D. Y. Patil Vidyapeeth, Pune, India
D. Anitha Kumari	School of Management Studies and Commerce, Vels Institute of Science, Technology and Advanced Studies (VISTAS), Pallavaram, Chennai-117, Tamil Nadu, India
Dharmendra Kumar	Faculty of Law, Shia P.G. College, Lucknow, U.P., India
Dheeraj Nim	Faculty of Management, Oriental University, Indore, India
Dipti Malpani	Faculty of Management, Oriental University, Indore, India
Jayesh Katkar	Zeal Institute of Business Administration, Computer Application & Research, Pune, India
Khursheed Jamil	Unity PG College, University Of Lucknow, Lucknow, U.P., India
Lambodar Saha	Lexicon Management Institute of Leadership and Excellence, Pune, India
Madhavi Udaybhan Shamkuwar	Zeal Institute of Business Administration, Computer Application & Research, Pune, India
Priya Jindal	Chitkara Business School, Chitkara University, Punjab, India
Rajasmita Panda	School of Computer Studies, Sri Balaji University, Pune, India
Renji George Amballoor	Directorate of Higher Education, Alto-Porvorim, Goa, India
Renuka Deshmukh	School of Business, Dr. Vishwanath Karad MIT World Peace University, Pune, India
Rishikaysh Kaakandikar	SaiBalaji International Institute of Management Sciences, Pune, India
Sarita Rana	Acropolis Faculty of Management and Research, Indore, India
Shankar B. Naik	Directorate of Higher Education, Alto-Porvorim, Goa, India
Shraddha Vijay Sinkar	SaiBalaji International Institute of Management Sciences, Pune, India
Sirshananda Panda	Alard Institute of Management Sciences, Pune, India
Sneha Raghuvanshi	Department of Management Studies, Medi Caps University, Indore, India
Sphurti Birajdar	Lexicon Management Institute of Leadership and Excellence, Pune, India
Supriya Pathak	Faculty of Management, Oriental University, Indore, India
Sushma Sharma	Department of Management Studies, SRM University, Delhi-NCR, Sonepat, India

T. Shalini Priyadharshini School of Management Studies and Commerce, Vels Institute of Science, Technology and Advanced Studies (VISTAS), Pallavaram, Chennai-117, Tamil Nadu, India

Yashashvi Jaiswal Department of Management Studies, Medi Caps University, Indore, India

CHAPTER 1

The Skill Revolution: Building a Resilient and Adaptive Workforce

Lambodar Saha[1], **Babasaheb Jadhav**[2,*], **Sphurti Birajdar**[1] and **Renuka Deshmukh**[3]

[1] *Lexicon Management Institute of Leadership and Excellence, Pune, India*

[2] *Dr. D. Y. Patil Vidyapeeth, Global Business School & Research Centre, Pune, India*

[3] *School of Business, Dr. Vishwanath Karad MIT World Peace University, Pune, India*

Abstract: AI is changing the global economy by altering job types and the skills that people need. This topic explores how AI and new ideas are changing our world and provides advice on preparing for these changes. The rapid growth of AI technology has already changed jobs, created new opportunities, and presented challenges for skill development. Tools like machine learning, natural language processing, and automation are changing how we work and creating a need for new skills. This needs ongoing support and skill development.

Good planning and ongoing learning are essential for keeping up with changes caused by AI. The talk discusses ways to improve AI skills, develop a lifelong learning habit, and adapt to new challenges. It highlights the role of strong education systems, job training programs, and development activities in preparing people for future jobs. Essential skills and good behavior are highlighted as key to financial success. There's a need for teachers, lawmakers, and business leaders to work together to create strong learning environments.

The text explains how people and organizations adapt to these changes by focusing on continuous learning and skill development. It highlights the importance of education, business support, and government actions in helping people learn for a long time and preparing them for future job needs.

The goal is to show how AI has improved workforce insights, helping people and organizations understand and use data and tools to succeed. It highlights the importance of basic AI knowledge for enhancing skills, solving problems, organizing tasks, and working together. People and groups can remain effective and adaptable by preparing for changes and improving AI skills.

[*] **Corresponding author Babasaheb Jadhav:** Dr. D. Y. Patil Vidyapeeth, Global Business School & Research Centre, Pune, India; E-mail: babasaheb.jadhav@dpu.edu.in

Rishikaysh Kaakandikar, Keshav Kaushik, Supriya Pathakm Madhavi Shamkuwar & Vivek Mishra (Eds.)

The writing survey examines how AI and machines are improving, how they are used in businesses today, and what we expect for the future. It examines research papers, industry reports, and books to explore how AI affects understanding and social progress.

The text examines how people who work with information can improve, identifies essential skills, and suggests how to develop those skills actively. It encourages exploring fields like data science and AI research and highlights that advanced AI skills can improve job opportunities and career growth. Successfully using AI can make tasks easier, provide new solutions, and assist in making decisions.

Essential insights and working together help develop creativity and basic thinking skills for understanding new problems. A promise to ongoing learning helps workers stay updated with the latest technology and skills. Being involved in AI groups and industry events helps improve skills, while management and business training support company growth and job creation. People and companies can encourage progress, development, and positive societal changes by understanding AI as a tool and learning valuable skills.

Keywords: Artificial intelligence, Adaptive workforce, Digital transformation, Future skills, Lifelong learning, Resilient workforce, Skill development, Skill revolution, Upskilling.

INTRODUCTION

Upskilling for the Age of AI Automation

In the age of the AI revolution, upskilling has become more than just a trendy term; it's a critical strategy for individuals and organizations aiming to succeed in the rapidly evolving landscape of artificial intelligence. As AI technology continues transforming industries and job roles, the demand for a workforce with the necessary skills has never been more urgent. In the context of the AI revolution, upskilling involves gaining new competencies or enhancing existing ones to meet the requirements of an AI-driven economy.

The rapidly changing landscape of the 21st century is undergoing a profound transformation with the integration of AI, which is fundamentally altering industries and redefining job roles.

The Skill Revolution

The world is rapidly digitizing, and AI skill requirements are also growing. A recent report by the OECD predicts that by 2030, 14% of jobs will likely be automated. The numbers vary significantly among countries and across different occupations. For example, postal and courier services, land transport, and food services face a higher risk of automation than CEOs and senior officials. The integration of AI into workplaces is, however, a double-edged sword.

AI and Future Work

It is undisputable that AI and future work are deeply interconnected. The impact of AI and automation on the workforce presents challenges and opportunities. AI is transforming industries by advancing innovation, improving efficiency, and creating new job categories, while automation has streamlined processes, reduced errors, and allowed workers to focus on complex tasks.

Anticipating Future Skill Requirements

Based on this research, the following conceptual framework suggests a proactive and collaborative approach to meeting skill needs, considering trends in automation and AI. This framework emphasizes the importance of close collaboration among all stakeholders and outlines the four key skills development responsibilities.

LITERATURE REVIEW

The writing audit indicates that requests for future abilities are estimated to change across different industries, with predictions suggesting varying change levels. This change is also apparent in Ireland. Both Deshpande and others. In 2021, Gallagher (2019) pointed out that robots and AI will make it more important to focus on human skills, especially soft skills. Most people agreed with this idea. Even though there have been many discussions about whether to focus more on soft or hard skills (Rotatori *et al.*, 2021), most people agreed with Zahidi *et al.* In 2020, there was a more significant focus on critical soft skills.

Teachers (Smith and White, 2020) and policymakers (Jakabsone, 2021) believe that teaching STEM subjects is essential early on. However, group members didn't see these skills as more important than soft skills for a long time. This view is appealing because jobs in STEM fields, like cybersecurity experts, data scientists, and engineers, are some of the hardest to fill. This suggests that the focus on STEM skills might be too much, or there could be a belief that, as technology advances, hard skills will become simpler to learn through large training programs from people, companies, and governments. In contrast, human soft skills will still be very valuable.

According to the rules by Santos and Reinhard (2012) and a report by the ILO (2020), participants emphasized the importance of better teamwork between the government, schools, professional groups, and businesses. This collaboration is needed to make education more effective and meet the needs of the manufacturing industry.

Shah and Burke (2003) suggested that training programs funded by free money work best when there is a clear understanding of what employers need in a specific industry. This might explain why many people praised Skillnet, a government group that works closely with businesses to create tailored training programs.

Many people believed encouraging businesses was crucial for helping employees learn new skills. Even so, it's well-known that many employers don't very often use available incentives, like giving gifts. People praised the current Apprenticeship Program and suggested expanding it to include more knowledge-based jobs as an option for college courses (McGrath, 2020).

As jobs became more flexible and more rigidly defined, some people argued that workers would need to develop both soft and hard skills to succeed in the future workplace. Teng and others back this view. In 2019 and again in 2022, Weritz found that Industry 4.0 will need a combination of both categories of skills. People said we should focus on hiring individuals with the right mindset who can learn quickly. After that, we should train them in the critical skills needed to fill gaps.

Everyone agreed that robots and AI would change the workplace; some expected these changes to be significant. However, some people thought that the impact of Industry 4.0 might be overstated. They compared this with recent findings and disagreed with the predictions that AI will replace many jobs for humans. This opinion exists because Ireland relies on the information economy and the essential skills that come with it. SMEs are believed to be slower in adopting new technologies. Those who are using AI are doing so in a basic way. This leads many people to think that the changes from Industry 4.0 will happen gradually, giving companies time to adjust.

Most members agreed that organizations in Ireland are struggling to meet recent skill needs, specifically small and medium-sized businesses (SMEs), which must compete with larger multinational companies. Some people pointed out the importance of better ways to deal with this issue (Rotatori *et al.*, 2021; Tyszko and Squanders, 2015). They suggested that COVID-19 has allowed businesses to attract more potential employees by offering flexible and remote work options. Members also emphasized the importance of hiring people who are adaptable and willing to learn rather than just focusing on specific technical skills that the company can teach (Shipley and McGowan, 2020; Lorri, 2013).

Instead of contesting for limited skills, we should share skills within the organization and with outside partners (Baldassari and Roux, 2017). It was also found that groups should be more willing to work together from a distance to create bigger groups of talent for hiring (Watchman and Kramer, 2011). Since keeping good workers is challenging, most members pointed out that training workers from within the company could be a good idea. This way, companies are better prepared for changes in the job market (Wilkinson *et al.*, 2021).

UPSKILLING FOR THE AGE OF AI AUTOMATION

The Changing Work of Nature

AI's rise dramatically reshapes work environments. Routine tasks are automated faster than ever, freeing individuals and organizations to focus on the most sophisticated and innovative endeavors. This shift highlights the importance of cultivating skills AI cannot emulate, such as critical thinking, emotional intelligence, creativity, and adaptability.

The Importance of Upskilling

a) Staying Relevant: As the landscape evolves, those prioritizing continuous learning are better equipped to remain competitive. b) Boosting Employability: In today's job market, where adaptability and responsibility to growth are highly valued, upskilling becomes a vital asset. c) Securing Future Careers: Upskilling is a forward-looking strategy designed to secure one's career against the unpredictable changes brought by AI.

Key Skills for the AI Era

- Critical Thinking and Problem Solving: While AI excels at processing data, human insight is crucial for complex decision-making and problem-solving.
- Creativity: Creativity is a distinctly human capability that AI struggles to match.
- Adaptability: As the workplace evolves at an unprecedented pace, adapting to new technologies and processes becomes increasingly critical.
- Emotional Intelligence: In roles that involve interpersonal interaction, where AI currently falls short, emotional intelligence is key.

The Role of Employers

- **Investing in Workforce Development:** Leading organizations recognize the value of upskilling and spend it on their employees' growth.
- **Providing Accessible Learning Resources:** Companies play a crucial role in supporting upskilling by offering accessible learning opportunities.

- **Cultivating a Learning Culture:** Promoting a culture that prioritizes continuous learning is essential.
- **Integrating Numly for Effective Upskilling:** Platforms like Numly are instrumental in developing the learning experience in the context of upskilling for the AI era. Numly provides a structured framework for skill enhancement and promotes the value of endless learning within organizations.

Equipping the Workforce with AI Reskills

Global industry transformation is occurring because of a new industrial revolution driven by emerging technologies and advances in artificial intelligence. It is now essential for professionals to be flexible and constantly learning to succeed in the digital age. Therefore, reskilling is critical for people and businesses to succeed in this new environment and stay competitive. At Trivaeo, we recognize the value of staff upskilling and reskilling; because of this, during the past three years, our team members have studied a wide range of topics, all of which have been fully funded by the company. These topics include design thinking, executive strategy, chief technology officer programs, artificial intelligence, advanced leadership and management, and non-executive director certifications.

Key Stakeholders in the Reskilling Revolution

This investment is about more than adjusting to change; it's about protecting and investing in your workforce's future. Companies can equip their workforce with the technical, creative, and digital skills necessary for success in an AI-driven environment by working with universities and encouraging a learning culture. Retaining talent through organizational changes requires providing avenues for career transfers and encouraging staff members to advance their abilities. The AI Revolution highlights people's need to adapt and expand their skill set. Employees whose jobs are automated ought to embrace digital literacy, data science, and artificial intelligence courses. Proactively improving one's abilities leads to job advancement and helps achieve the larger objective of developing a competitive workforce led by innovation.

Economic Competitiveness

Nations and corporations must make labor investments if they want to stay competitive. Your workers can be upskilled and reskilled to guarantee that no one gets left behind and that qualified candidates fill positions. This strategy's cascading consequence is innovation-driven economic growth. Additionally, reskilling lowers the unemployment rate and boosts productivity because employees are guaranteed to possess all the skills needed to do their jobs well.

Doing this contributes to developing a strong economy that is also very resilient to changes in the world market.

Job Security and Career Growth

Employees becoming more worried about losing their jobs due to automation and artificial intelligence can increase their chances of advancement and secure employment by reskilling their personnel. Employers benefit from having employees with updated skills since they will naturally transfer to other positions within the company rather than taking their talent elsewhere. Regular reskilling also lets your employees and company stay updated with rapid technological advancements. By being more proactive rather than reactive, it is possible to avoid the need for layoffs and to open new and higher-level positions that will provide stability and long-term career progression.

Bridging the Skills Gap

The current labor market cannot close the skills gap caused by the digital and AI revolution. Reskilling fills the void, restoring a worker's relevance, considering market demand. This category also includes software development, cybersecurity, and data science. As a result, the workforce would be prepared with cutting-edge industry trends and current skills. Regular curriculum updates and new learning environments guarantee that workers in a changing labor market are adaptable and versatile.

Reskilling Challenges

This chapter examines the significance of reskilling for businesses and employees. I also discuss the difficulties of reskilling and the essential tactics for program implementation so you can use digital and Artificial Intelligence (AI) in your work or business.

Determining the Skills Needs

Determining with reasonable accuracy which abilities will be relevant in the future is a significant challenge in reskilling. This necessitates the government, business, and academic communities to share their thoughts and have foresight regarding the technological developments that will likely alter the labor market. This joint endeavor consists of data analysis on developing technology domains, comprehending industry requirements, and forecasting potential future employment market conditions. Stakeholders may benefit from insights from other industries to build up targeted reskilling, ensuring the necessary skills are included to leave the workforce ready for challenges and change.

Case Studies and Real-Life Examples

Singapore's Skills Future initiative

Singapore's Skills Future is a comprehensive lifelong learning and skills development program. It provides a broader selection of courses, financial aid, and career counseling services to improve competencies and stay competitive in an evolving labor market. It has effectively established a culture of lifelong learning for Singaporeans' employability.

The program IBM Skills Build

Through IBM's Skills Build program, members of underrepresented populations can access free online resources for digital skill training. The initiative offers personally tailored training to reposition people for new jobs currently in demand within the information economy in collaboration with organizations, municipalities, and schools. This is where public-private reskilling initiatives shine.

The Dual Education System in Germany

The dual education system in Germany combines traditional classroom instruction with on-the-job training for practical skills. This approach has successfully educated youth for careers in engineering and manufacturing, which are in high demand. The dual education system is one functional reskilling element that might function when theoretical and practical elements are combined.

Workforces must be reskilled for the digital and artificial intelligence era to minimize the skill gap, maintain job security, and remain competitive. These fundamental duties are all beset by several formidable obstacles. Public-private collaborations, lifelong learning, specialized programs for acquiring new skills, and utilizing technological advancements are still viable means of contributing to the advancement of society. The advent of online educational platforms such as LinkedIn Learning, Coursera, and YouTube has made reskilling more accessible and practical. Proactive investors and students find it highly appealing when companies, governments, and educational institutions fund reskilling programs that prepare workers for the digital transformation of the workforce of the future.

THE SKILL REVOLUTION

The Rise of AI is prepared to create avenues for the emergence of new industries, much like the internet led to the birth of e-commerce and social media. We may see the advent of industries that are currently beyond our imagination. It's essential to cultivate a climate that nurtures innovation and entrepreneurship, enabling these industries to thrive and generate employment opportunities.

Beyond Job Displacement: AI's Role in Enhancing Productivity and Growth

AI's economic impact stretches beyond job creation. By automating repetitive tasks, AI can significantly improve productivity, enabling businesses to deliver more goods and services with fewer resources. This boost in efficiency drives economic growth, which, in turn, can create new jobs. Moreover, AI can potentially convert sectors like healthcare and finance, leading to better services and opening doors for new wealth-generating opportunities.

The Human Advantage: Essential Skills in an AI-Driven Future

While AI offers tremendous possibilities, it's important to recognize that human roles will remain vital in the workforce. Skills distinguishing us from machines, such as critical thinking, creativity, problem-solving, and emotional intelligence, will be invaluable in AI. There should be a rigorous attempt to equip the workforce with these uniquely human capabilities. Educational systems must promote critical thinking, communication, collaboration, and adaptability. Furthermore, fostering a culture of lifelong learning will be crucial, enabling workers to continually acquire new skills and stay competitive in the evolving job market.

The Role of Policy and Social Safety Nets

Transitioning to an AI-driven economy will present challenges. Even temporary job displacement can lead to economic difficulties for individuals and families. Governments and policymakers must take a proactive approach by investing in retraining initiatives, strengthening social safety nets, and enacting policies that support workers during these transitions. Additionally, as AI automates more tasks and potentially reduces the need for human beings, discussions around income inequality and the idea of universal basic income may gain traction. Ensuring everyone benefits from AI's economic gains will be vital to maintaining social stability and fostering a future where all can prosper.

GLOBAL POLICY RESPONSES TO AI-DRIVEN WORKFORCE TRANSFORMATION

As Artificial Intelligence (AI) and automation reshape the global labor market, national governments and supranational bodies have launched ambitious initiatives to prepare their workforces for the evolving digital economy. These policy responses focus on digital literacy, lifelong learning, and public-private partnerships to upskill citizens and close emerging skill gaps.

India: Skill India Mission and FutureSkills Prime

India's Skill India Mission, launched in 2015, seeks to train over 400 million individuals in market-relevant skills by 2025. It encompasses a wide range of programs, including the Pradhan Mantri Kaushal Vikas Yojana (PMKVY) and FutureSkills Prime, the latter being a joint initiative by NASSCOM and the Ministry of Electronics and Information Technology. FutureSkills Prime promotes digital fluency and professional readiness in areas like AI, cloud computing, and cybersecurity, ensuring alignment with industry demands and job market trends (NASSCOM, 2021).

European Union: Digital Skills and Jobs Coalition

The European Union has spearheaded the Digital Skills and Jobs Coalition under its broader Digital Europe Programme (2021–2027). This initiative aims to foster a digitally competent society by coordinating efforts across member states, businesses, and education sectors. The DigComp Framework, developed by the European Commission, offers a structured methodology to assess and improve digital competences across Europe (European Commission, 2022).

Singapore: SkillsFuture Initiative

Singapore's SkillsFuture initiative is a global exemplar in creating a culture of lifelong learning. Every Singaporean above the age of 25 receives training credits to pursue industry-relevant education, including digital and AI-focused courses. The government collaborates with global technology companies and institutes of higher learning to align the program with emerging job roles in data analytics, machine learning, and digital marketing (SkillsFuture Singapore, 2023).

United States: National AI Initiative and Workforce Programs

The National AI Initiative Act of 2020 lays the foundation for coordinated federal investments in AI education and workforce development. It funds AI research and supports educational programs through community colleges and universities, while also backing the expansion of apprenticeship models and public-private partnerships in AI workforce readiness (U.S. Government, 2021).

Germany: Digitalpakt Schule and Vocational Education Reforms

Germany continues to modernize its robust dual vocational training model through DigitalPakt Schule, which allocates funding to upgrade digital infrastructure in schools. Reforms in vocational education now embed digital and AI-related curricula, preparing young learners and apprentices for the technological demands of Industry 4.0 (German Federal Ministry of Education and Research, 2022).

International Organizations: OECD and UNESCO Initiatives

The OECD and UNESCO play pivotal roles in setting global benchmarks and supporting national policy frameworks. The OECD's Skills Outlook reports provide evidence-based recommendations for AI-aligned education systems, while UNESCO emphasizes equitable access to digital learning and capacity-building in developing nations (OECD, 2023; UNESCO, 2021).

HOW TO RESKILL YOUR WORKFORCE'S IMAGE OF AI

The beginning of the AI era is now evident to most of us. AI is set to revolutionize nearly every facet of business, from data analytics and product development to customer experience and beyond. A key question is how AI will impact businesses and employees. According to Raffaella Sadun from Harvard Business School, the outcome hinges on how effectively an organization integrates new technologies with a comprehensive corporate strategy and the professional development of individual employees. Focusing on one without the other risks job losses and fewer opportunities for everyone. Fortunately, as Sadun notes, we are still in the initial stages of AI adoption, meaning the future is still open to influence. Leaders don't need to grasp every detail of AI but should focus on identifying the best applications for their business and communicating a clear strategy for reskilling their teams. The skills-first approach fosters a more adaptable workforce.

Getting the 'Skills' Transformation Right

A skills-first approach starts with rapid-fire pool planning to

- Understand the demand for specific skills presently and in the future to deliver on business objects.
- Assess force, uncovering what chops presently live in the association (this should include workers and contractors, as well as seeker and alums gift pools)
- Dissect skill gaps and develop a plan to close them.

A well-defined skill-based job is a critical stepping stone to effective pool planning. Most associations struggle to keep skills data up to date and accessible. For most businesses, disconnected systems do not update as people develop and add to their skill sets. This is where an AI skills conclusion platform can help by creating a clear and harmonious chops taxonomy specific to each association that adapts automatically as the business evolves.

Using AI to Map Skills

Anticipating the future condition of demand and modeling the internal and external market supply in the skill supply model are essential components of developing a strategy to close skills gaps. You can create internal mobility, employee development, and hiring strategies that are more focused explicitly by knowing the skills you need and those you already possess. One of the most significant ways to ensure your company can adjust to quickly changing talent demands is to have a talent pipeline full of engaged prospects that recruiters can interact with at any time and have a clear perspective of their skills. Going one step further, you may develop a "skills adjacency map" with an AI-powered skills board showing you how close one talent is to other activities.

Skills Development for Ongoing Evolution

If you want to create an agile and resilient business in a costly labor market, staff retention and internal skill development are more crucial than ever. The following should be included in large-scale skill development programs:

- Well-defined development routes grounded in job architecture and skill adjacency insights; Establishment of organizational frameworks for education, capacity-building, and guidance
- Strict procedures for monitoring these programs' effects.

The most recent assessment of the skills gaps you urgently need to fill should always serve as the foundation for recommendations on education, job rotations, and even new mentors. If skills development is prioritized, the workforce will always change, which is crucial in today's quickly evolving business climate and among the greatest support methods.

Developing Skills in Intelligence

By putting skills first, you develop the intelligence required to make more informed decisions about personnel and design adaptable workforce strategies. You may evaluate talent availability internally with the help of your organization's skills taxonomy and job architecture, and you can supplement this with market-wide insights. With the correct technology, you may learn things like a person's propensity to respond to a job posting or move. Find out which skills are the most important, use filters to slice your data, and use AI on the data to draw conclusions and get insights. Examples of these are the skills typically associated with other skills, unusual skill pairings, and the proportion of people with skills compared to the demands in your future.

Equitable, Impartial, Fair, and Transparent Gathering of Skill Data

Decisions made because of systematic biases in data collection and evaluation will continue to exist disparities. Organizations should commit to clear, consistent, and inclusive procedures for skill assessment and to continuously analyze and improve collecting techniques to ensure equity if they are to fully realize the potential of skill-based talent management. Leveling the playing field promotes meritocracy and helps skill-based approaches reach their full potential.

Managerial and Organizational Faith in the Skill Data

The effectiveness of skill-based talent management depends on managerial and organizational trust. Without confidence in the impartiality and accuracy of skill data, people are prone to turn back to well-known (and erroneous) criteria like schooling and personal relationships. These can reinforce prejudices and obstruct the integration of varied skills, resulting in an echo chamber instead of a space for novel ideas. Talent management focused on skills offers a chance to improve equity and productivity at work. An atmosphere that prioritizes merit and inclusivity over connections or background information is achieved by focusing on employees' skills and contributions. However, businesses must build trust and ensure fair data-gathering procedures to benefit fully. Only then can they transform into the way they are.

THE AI REVOLUTION: UPSKILLING FOR THE AUTOMATED FUTURE 2024

You must improve your skills, get better tools, take on more tasks, and make good connections. Now is the time to change from being easy to replace to being very valuable. Time is moving, and the future won't wait for you to prepare.

The AI Revolution In 2024

It's time to get better at what you do. Try free coding classes, study statistics, and improve your teamwork skills. Learning new things will be important. As AI handles easy jobs, people will focus on creativity, thinking deeply, and solving complex problems. Some jobs will disappear, of course. Sorry, telemarketers and data entry workers, your jobs won't last longer. But there will be many new jobs, like AI ethics experts, data analysts, and robot programmers. Some jobs are safe for now. AI still struggles with tasks that require emotions, quick decision-making, and skilled hand movements.

The main thing is to view AI as a helpful tool, not something to be afraid of. If you learn to use AI tools, you'll open more opportunities for yourself. Let AI take care of the dull jobs so you can focus on more important things.

How to Prepare for the Rise of AI

Are robots going to take your job? Don't worry too much right now. Even though AI can change many industries, humans still have some unique skills.

Upskill Like Your Job Depends on It

Learn as much as possible about how AI and machines change your work. Take extra classes to understand the latest tools and technology in your job field. Practice skills that are unique to people, like being creative, knowing how others feel, and solving challenging problems.

Flex Your Leadership Muscles

As AI changes the path we work, new jobs will be created in managing and leading people. Make yourself a better job candidate by helping with extra tasks at work. Assist your coworkers. Improve your thinking for planning and making decisions.

THE SKILL REVOLUTION: PREPARING FOR THE FUTURE

AI has developed a lot recently, including robots, understanding human speech, and learning from information. AI is used more often in different areas, helping people do things better and changing their work. For example, chatbots, virtual helpers, driverless buses, and smart factories exist. AI can do tasks independently, look at data, and give helpful information. This helps people work better and be more creative in many fields. This is why it can change things a lot. The AI revolution has brought many changes.

AI is increasingly used in many fields, such as healthcare, finance, robots, data analysis, speech understanding, self-driving buses, accuracy, communication, technology growth, and quick problem-solving. AI has changed jobs and the amount of work people must do, bringing benefits and difficulties. The development of AI has led to more robots being used, which makes everyday tasks better and more accurate in many areas, like business, making things, moving goods, money management, and healthcare. A report from the World Economic Forum says that by 2025, AI could create 12 million more jobs worldwide than it will eliminate.

WHICH JOBS MAY AI DISRUPT BY 2030?

There are both chances to do well and problems to face. For example, more people began using home phones when cell phones became popular. When online booking systems began, ticket sellers lost their jobs. The arrival of ATMs reduced the number of bank tellers employed. As AI grows fast, it might create many job issues. This means that current problems with AI are essential. We need to be concerned about losing jobs because of AI. These are the jobs that will likely be affected by significant changes from AI. To prepare for the future, we need to understand which jobs might be replaced by robots. These include jobs like data entry, factory work, customer service, and other jobs with similar tasks.

Eventually, some jobs may go away and be replaced by AI and machines. Robots are now doing jobs on assembly lines that people used to do. They are usually right, which makes them reliable, but this can lead to significant issues in people's lives. People will do less physical work because of improvements in AI technology. This might lead to people losing their jobs and changing how companies use technology. People who don't seize chances and instead fight against difficulties might find it tougher to get good-paying jobs today. Using AI will run new jobs, but it will also take away some current jobs that can be done by similar technology.

FUTURE WORK

Upskilling and Reskilling for an Automated Future

Are robots going to take your job? Not if you don't allow it. The secret to keeping your job is effortless: make yourself someone no one can replace.

Develop Human Skills

Focus on building the qualities that make us human, like being creative, understanding emotions, and making wise choices. Computers can't be like that. Improve your people skills and welcome the more flexible aspects of your job. While the robots take care of everyday tasks, you can focus on essential things that need a personal touch, like leading teams, fixing problems, and calming unhappy customers.

Learn to Code

Coding might not be your thing, but learning technical skills is essential. Try looking at data, online marketing, or designing user experiences. The more you know about technology, the better you'll use it to your advantage instead of trying to fight against it.

Take on Responsibility

Find ways to take on responsibility, especially in tasks that highlight what people do best. Focus on important, overall tasks where your understanding of emotions and ability to make choices will be helpful. Taking care of people and projects is a good choice.

Keep Learning

What you know and can do now might not be useful in five years. Keep learning regularly to ensure your skills stay current with new technology and business methods. Enroll in online classes, get guidance, attend industry events, and follow experts in your area.

Embracing AI Tools at Work

Robots will take your job. The good news is that they likely only want the tedious tasks you don't like doing, like expense reports and timesheets. The bad news is that you could lose your job if you don't get along with the new machines at work.

Buddy up to the Bots

Getting along with the robots is essential to keeping your job safe during AI. Take them out for drinks, tell them you like their shiny metal bodies, and ask them many questions about machine learning. You should get used to working with AI systems and software.

Learn the Lingo

Add "good at chatting in Python" and "skilled in SQL" to your resume. While the robots take care of the regular tasks, you should focus on improving your skills. Concentrate on building skills AI finds hard, such as talking to others, working together, and solving complex problems. Take online classes that go well with AI to improve your tech skills, like looking at data, statistics, and machine learning.

Make Yourself Indispensable

The growth of AI is going to happen, but if you plan smartly, you can avoid getting left behind as technology moves forward. Please familiarize yourself with the bots, understand how they communicate, and focus on being a team member that no AI can take the place of. If you have some luck and skill, you and the robots can get along well and even help each other work better. In the future, many machines will be doing things, but people are still important.

Protecting Your Job by Focusing on Human Skills

The important part is building the skills that AI has a hard time with. Skills like being creative, thinking carefully, and understanding feelings. First, learn new skills because your job relies on them. How can you use new AI tools to help you instead of letting them take your job? Sign up for online courses to improve skills that are harder for computers, like solving challenging problems or making essential choices. The more skills you have, the less likely people will see you as someone who can be easily replaced.

AI AND SKILLS OF THE FUTURE

Every day, there's new news about how AI is changing everything. In the workplace, technology with AI will change our jobs and how we do them.

We requested our research associate, Pearson, to examine six countries closely: Germany, the USA, Australia, the UK, Japan, and India. We wanted them to determine what essential job skills people need in five years.

We wanted to learn how AI automation will change jobs in different fields and find out what business leaders and their workers can do now to get ready.

Shaping the Workforce of the Future

The research shows that as AI takes over machine tasks, people don't have to work like machines as much. Essential skills like talking to others, working together, being creative, and thinking critically will be the key tech skills wanted in the future.

Fortunately, these skills can be transferred from jobs that might be impacted by AI to future tech jobs. The study showed that, with proper training, administrative assistants in the U.S. can be great choices to fill the increasing need for help desk workers.

This is a positive thing. People weren't meant to copy and paste information into spreadsheets all day. However, we must take careful steps to change the future workforce.

Closing Tech Talent Opening Gap

IDC says that 90% of companies worldwide are missing a lot of technology help, which could cost them $6.5 trillion by 2025. There aren't enough tech workers from usual job backgrounds to meet the need. What seems like a lack of tech skills is an opportunity, and AI is helping business leaders fix it. AI is making tech jobs easier by handling more specific tasks. People with development skills or computer skills can get these jobs more easily. Business leaders must look at who they hire and how they do it. They need to train their workers to fill the future tech jobs. Programs like Rise with ServiceNow help people whom the tech industry has often ignored learn essential skills. We aim to train 1 million people on the Now Platform by the end of 2024. We want to help them build successful careers and support our guests and friends in finding what they need to succeed in the digital economy. Even though the messages about AI might seem scary, the information shows a different picture. It's time for everyone to review their position in the pool and get excited about the upcoming opportunities.

CONCLUSION

In short, learning new skills because of AI is not just about keeping up with technology; it's also a smart way to improve yourself and your job prospects. As AI plays a more significant role in our work, the value of skills only humans can do will increase. Being eager to keep learning and adjusting is essential for success in the exciting and changing jobs influenced by artificial intelligence. It is not just about using new technology; it's about using it to achieve more. Platforms like Numly provide helpful tools to make exploring this exciting future easier and more effective.

Contact us to learn how our AI coaching platform can help you develop better leaders. These leaders can create strong teams that deal with today's changing work world with a better understanding, confidence, and clarity. Hey, robots are going to replace your job. The AI train has begun its journey, and we can't stop it now. Sitting around and feeling sad won't make things better. You should do things to keep your job safe for the future.

Learn new things, adapt to changes, and stay with technology, or you might be left behind. Robots can do typical jobs but will never be as imaginative, understanding, or daring as people. So, clean the Cheetos dust off your fingers, put down the game controller, and begin practicing the skills you need to do well in a world with more machines. The robots want to take your job, but you'll have to fight hard to keep it. Now, go out and show me what people can achieve.

The growth of AI represents an essential change in human history. Although many people fear losing their jobs, the story about AI doesn't have to be scary. AI can be a helpful way to make money, create new jobs, and make our lives better. To make this change successful, we must plan carefully, spend money on education and training, and develop programs that ensure everyone benefits from AI equally. If we accept a future where people and AI help each other, we can begin a new era of success and happiness for everyone.

REFERENCES

Baldassari, P., Roux, J.D. (2017). Industry 4.0: Preparing for the future of work. *People and Strategy, 40*(3), 20-24.

Bruun, E.P.G., Duka, A. (2018). Artificial intelligence, jobs and the future of work: Racing with the machines. *Basic Income Studies, 13*(2), 1-15.
[http://dx.doi.org/10.1515/bis-2018-0018]

Brunello, G., Wruuck, P. (2021). Skill shortages and skill mismatch: A review of the literature. *J. Econ. Surv., 35*(4), 1145-1167.
[http://dx.doi.org/10.1111/joes.12424]

Brynjolfsson, E., McAfee, A. (2014). *The second machine age: Work, progress, and prosperity in a time of brilliant technologies..* W.W. Norton & Company.

Cutler-Knowles, A., Lewis, H. (2018). Talent for survival: Essential skills for humans working in the machine age. *Deloitte.*

Calonge, D.S., Shah, M.A., Riggs, K., Connor, M. (2019). MOOCs and upskilling in australia: A qualitative literature study. *Cogent Education, 6*(1), 1-19.
[http://dx.doi.org/10.1080/2331186X.2019.1634256]

Collings, D., McMackin, J. (2019). Enabling the workforce of the future: The role of learning and development. *Skillnet Ireland.*

Chuang, S. (2021). An empirical study of displaceable job skills in the age of robots. *European J. Train. Dev., 45*(6/7), 617-632.
[http://dx.doi.org/10.1108/EJTD-10-2019-0183]

Deshpande, A. (2021). Improving working conditions using artificial intelligence. *Policy Department for Economic, Scientific, and Quality of Life Policies.*

Frey, C.B., Osborne, M.A. (2017). The future of employment: How susceptible are jobs to computerisation? *Technol. Forecast. Soc. Change, 114*(1), 254-280.
[http://dx.doi.org/10.1016/j.techfore.2016.08.019]

Grant, K., Venz, L., Bottcher, K. (2020). The dark side of digitalization? A diary study of the impact of digital technologies on work intensification, work-life balance, and wellbeing. *CIPD Applied Research Conference, Dublin.*

Georgieff, A., Hyee, R. (2022). Artificial intelligence and employment: New cross-country evidence. *Frontiers in Artificial Intelligence, 5*, 832736.
[http://dx.doi.org/10.3389/frai.2022.832736] [PMID: 35620279]

Huang, M.H., Rust, R.T. (2018). Artificial intelligence in service. *J. Serv. Res., 21*(2), 155-172.
[http://dx.doi.org/10.1177/1094670517752459]

Howard, J. (2019). Artificial intelligence: Implications for the future of work. *Am. J. Ind. Med., 62*(11), 917-926.
[http://dx.doi.org/10.1002/ajim.23037] [PMID: 31436850]

Jaiswal, A., Arun, C.J., Varma, A. (2021). Rebooting employees: Upskilling for AI in multinational corporations. *Int. J. Hum. Resour. Manage., 33*(6), 1179-1208.
[http://dx.doi.org/10.1080/09585192.2021.1891114]

Jadhav, B. (2023). *Workforce management system: Concepts, definitions, principles & implementation..* CRC Press.

Jadhav, B. (2023). *Industry revolution 4.0: Workforce competency models and designs..* CRC Press.

Jadhav, B. (2023). *AI-enabled approaches and models for workforce design through training systems for physically challenged people..* CRC Press.

Krzywdzinski, M. (2017). Automation, skill requirements and labour-use strategies: high-wage and low-wage approaches to high-tech manufacturing in the automotive industry. *New Technol. Work Employ., 32*(3), 247-267.
[http://dx.doi.org/10.1111/ntwe.12100]

Lane, M., Saint-Martin, A. (2021). The impact of artificial intelligence on the labor market: What do we know so far? *OECD Working Papers,* 256.
[http://dx.doi.org/10.1787/1815199X]

Makarius, E.E., Srinivasan, M. (2017). Addressing skills mismatch: Utilizing talent supply chain management to enhance collaboration between companies and talent suppliers. *Bus. Horiz., 60*(4), 495-505.
[http://dx.doi.org/10.1016/j.bushor.2017.03.007]

McDonald, M. (2020). *Digital Skills Insights 2020..* International Telecommunication Union.

McNaboe, J. (2021). *National Skills Bulletin 2021..* National Skills Council.

McGuinness, S., Pouliakas, K., Redmond, P. (2023). Skills-displacing technological change and its impact on jobs: challenging technological alarmism? *Econ. Innov. New Technol., 32*(3), 370-392.
[http://dx.doi.org/10.1080/10438599.2021.1919517]

Morandini, S., Fraboni, F., De Angelis, M., Puzzo, G., Giusino, D., Pietrantoni, L. (2023). The impact of artificial intelligence on workers' skills. *Inf. Sci., 26*, 039-068.
[http://dx.doi.org/10.28945/5078]

Pereira, V., Hadjielias, E., Christofi, M., Vrontis, D. (2023). A systematic literature review on the impact of artificial intelligence on workplace outcomes: A multi-process perspective. *Hum. Resour. Manage. Rev., 33*(1), 100857.
[http://dx.doi.org/10.1016/j.hrmr.2021.100857]

Rotatori, D., Lee, E.J., Sleeva, S. (2021). The evolution of the workforce during the fourth industrial revolution. *Hum. Resour. Dev. Int., 24*(1), 92-103.
[http://dx.doi.org/10.1080/13678868.2020.1767453]

Su, Z., Togay, G., Cote, A.M. (2021). AI: A destructive and yet creative force in the skilled labor market. *Hum. Resour. Dev. Int., 24*(3), 341-352.
[http://dx.doi.org/10.1080/13678868.2020.1818513]

Shipley, C., McGowan, H. (2020). *The adaptation advantage: Let go, learn fast, and thrive in the future of work..* John Wiley & Sons.

Teng, W., Ma, C., Pahlevansharif, S., Turner, J.J. (2019). Graduate readiness for the employment market of the 4th industrial revolution. *Educ. Train., 61*(5), 590-604.
[http://dx.doi.org/10.1108/ET-07-2018-0154]

Webb, M. (2020). The impact of artificial intelligence on the labor market. *Social Science Research Network.*
[http://dx.doi.org/10.2139/ssrn.3482150]

CHAPTER 2

The Future of Work: In-Demand Skills for the Next Decade

Rajasmita Panda[1,*] and **Sirshananda Panda**[2]

[1] *School of Computer Studies, Sri Balaji University, Pune, India*

[2] *Alard Institute of Management Sciences, Pune, India*

Abstract: The work landscape is evolving as we move into the 21st century. Driven by technological advancements, globalization, and shifting economic paradigms, the skills in high demand over the next decade significantly differ from those that dominated the previous eras. Understanding these trends is crucial for individuals and organizations aiming to thrive in the future economy. The work landscape is rapidly evolving, driven by technological advancements, globalization, and shifting societal needs. As we look ahead to the next decade, specific skills will become increasingly essential for individuals navigating this dynamic environment. This abstract explores the in-demand skills expected to shape the future of work, focusing on their significance, emergence, and implications for individuals and organizations.

Keywords: Adaptability, Digital literacy, Data science, Global awareness, Interpersonal skills, Project management, Sustainability.

INTRODUCTION

The work landscape is profoundly transforming as we advance into the next decade. Emerging technologies, economic shifts, and evolving societal expectations are reshaping industries and redefining the skills crucial for success in the future workplace. This introduction explores the critical in-demand skills expected to shape the future of work, highlighting their significance in navigating the complexities of a rapidly changing global economy.

Various experts and organizations have extensively analyzed the evolving landscape of work and the skills required for the future. McKinsey Global Institute (2023) highlights the need for adaptability and digital skills by 2030 as workplaces undergo significant transformations. Similarly, Manyika *et al.* (2017)

* **Corresponding author Rajasmita Panda:** School of Computer Studies, Sri Balaji University, Pune, India; E-mail: rsmita1024@gmail.com

Rishikaysh Kaakandikar, Keshav Kaushik, Supriya Pathakm Madhavi Shamkuwar & Vivek Mishra (Eds.)

emphasize the duality of automation, which is likely to create new jobs while displacing others, requiring workers to upskill rapidly. The World Economic Forum (2022) reiterates this by predicting shifts in demand for technological and social skills, suggesting a focus on lifelong learning to remain competitive. Brynjolfsson and McAfee (2014) explore the profound effects of technological advancements on productivity, economic disparity, and the nature of work in their analysis of the "Second Machine Age." Government efforts, such as those outlined by the United States Department of Labor (2022), underline the importance of workforce readiness and skills development policies to meet future demands. Resources like LinkedIn Learning (n.d.) also advocate for proactive upskilling in collaboration, leadership, and technology to thrive in the rapidly changing job market. These perspectives collectively underline the urgency of adapting to a dynamic workforce landscape shaped by automation and innovation.

KEY THEMES EXPLORED

- **Technological Advancements:** The rise of automation, artificial intelligence (AI), and machine learning is revolutionizing industries across sectors. Digital skills, including data analysis, coding, and cybersecurity, will be essential for leveraging technology to drive innovation and efficiency.
- **Adaptability and Agility:** With the pace of change accelerating, adaptability and agility are becoming core competencies. Individuals who can quickly learn new skills, pivot roles, and embrace change will thrive in dynamic work environments characterized by disruption and uncertainty.
- **Human-Centric Skills:** While technology plays a pivotal role, human-centric skills such as communication, empathy, and collaboration remain indispensable. These skills foster effective teamwork, leadership, and customer relations, enhancing organizational resilience and employee satisfaction.
- **Sustainability and Environmental Awareness:** Heightened awareness of environmental issues and sustainable practices reshapes corporate priorities. Skills related to sustainable development, environmental stewardship, and green technologies are increasingly valued as organizations strive for ethical and responsible business practices.
- **Critical Thinking and Problem-Solving:** In a complex and interconnected world, the ability to critically analyze information, make informed decisions, and solve complex problems creatively will be in high demand. These skills empower individuals to navigate ambiguity and drive innovation in their respective fields.
- **Global Competence:** As businesses operate globally, cultural competence and the ability to work across diverse cultural contexts are essential. Understanding global markets, respecting cultural differences, and adapting business strategies will facilitate international collaboration and expansion.

- **Lifelong Learning:** Continuous learning and upskilling will be imperative as industries evolve and new technologies emerge. Individuals committed to ongoing self-improvement and staying abreast of industry trends will remain competitive in the evolving job market.

This introduction sets the stage for exploring these themes in depth, highlighting their significance in shaping the future of work and guiding individuals and organizations toward strategic skill development. By understanding and preparing for these in-demand skills, individuals can position themselves for success and contribute to a resilient, innovative, and inclusive workforce in the years to come.

Digital Literacy and Technology Proficiency

Technology is at the forefront of future skill demands. Digital literacy is no longer optional but essential across all industries. This encompasses:

- **Basic Digital Skills:** Proficiency in using digital tools and platforms is foundational. This includes familiarity with operating systems, office suites, and basic troubleshooting.
- **Advanced IT Skills:** Expertise in software development, cybersecurity, and data analytics is becoming increasingly vital. Professionals in these areas must innovate, secure, and optimize digital infrastructures.
- **Emerging Technologies:** Knowledge of artificial intelligence (AI), machine learning, blockchain, and the Internet of Things (IoT) is highly sought after. These technologies drive major transformations across healthcare, finance, and manufacturing.

Data Science and Analytics

In an era where data is often referred to as the new oil, the ability to analyze and interpret large datasets is invaluable. Skills in data science enable organizations to make data-driven decisions, personalize customer experiences, and optimize operations. Key competencies include:

- **Statistical Analysis:** Understanding statistical methods to analyze and interpret data trends.
- **Machine Learning:** Developing algorithms that can learn and make predictions from data.
- **Data Visualization:** Presenting data clearly and compellingly to stakeholders through tools like Tableau and Power BI.

Adaptability and Continuous Learning

The rapid pace of change requires an adaptable workforce committed to lifelong learning. Skills in this category include:

- **Agility:** The ability to quickly pivot and adapt to new technologies and methodologies.
- **Resilience:** Coping with and recovering from challenges and setbacks.
- **Learning Agility:** Continuously acquiring new skills and knowledge, often through self-directed learning or formal education programs.

Interpersonal and Soft Skills

While technical skills are crucial, soft skills remain equally important. As automation and AI handle more routine tasks, human-centric skills gain prominence:

- **Communication:** Effective verbal and written communication is essential for collaboration and leadership.
- **Emotional Intelligence:** Understanding and managing one's emotions and those of others to foster a productive work environment.
- **Creativity and Innovation:** The ability to think outside the box and develop novel solutions to complex problems.

Sustainability and Environmental Awareness

As global focus shifts towards sustainability, skills in this area are increasingly valuable. This includes:

- **Environmental Science:** Understanding ecological principles and sustainability practices.
- **Green Technology:** Developing and implementing technologies that reduce environmental impact.
- **Corporate Social Responsibility (CSR):** Integrating social and environmental concerns into business operations and strategy.

Project Management

With projects becoming more complex and cross-functional, effective project management is critical. Skills in this domain include:

- **Agile Methodologies:** Familiarity with Agile, Scrum, and other iterative project management techniques.
- **Risk Management:** Identifying, assessing, and mitigating risks throughout the project lifecycle.
- **Resource Allocation:** Efficiently distributing resources to maximize project outcomes.

Global and Cultural Awareness

In a globalized economy, working effectively across cultures is a significant asset. This entails:

- **Cross-Cultural Communication:** Understanding and respecting cultural differences to collaborate effectively with global teams.
- **Foreign Languages:** Language proficiency in Mandarin, Spanish, or Arabic can be advantageous.
- **Global Market Insights:** Knowledge of international markets and global business trends.

The future of work demands a blend of technical proficiency, continuous learning, and strong interpersonal skills. As technology evolves and global challenges emerge, staying ahead in the workforce requires a proactive approach to skill development. By focusing on these in-demand skills, individuals can position themselves for success in the dynamic landscape of the future economy.

THE DETAILED EXPLANATION OF THE ABOVE THEMES IS BELOW

Digital Literacy and Technology Proficiency

Introduction

Digital literacy and technology proficiency are foundational skills crucial in the future workplace. As technology continues to advance and integrate into every aspect of our lives, proficiency in digital tools and understanding emerging technologies are no longer optional but essential. These skills enable individuals to navigate the digital landscape, leverage new tools for productivity, and stay competitive in the job market.

COMPONENTS OF DIGITAL LITERACY AND TECHNOLOGY PROFICIENCY

Basic Digital Skills

- **Proficiency with Digital Tools:** This includes using typical software applications such as word processors, spreadsheets, and presentation tools. Proficiency in these areas is necessary for day-to-day tasks in most jobs.
- **Internet Navigation:** Understanding how to effectively search for information online, evaluate the credibility of sources, and use various online services and platforms.
- **Communication Tools:** Competence in using email, messaging apps, and video conferencing tools is essential for collaboration in a digital workplace.

Advanced IT Skills

- **Software Development:** Knowledge of programming languages (such as Python, Java, or C++) and software development methodologies enables individuals to create and maintain software applications.
- **Cybersecurity:** Understanding how to protect systems and data from cyber threats is crucial as cyberattacks become more sophisticated.
- **Data Analytics:** The ability to analyze and interpret data to make informed decisions. Skills in using tools like SQL, R, or Python for data analysis are highly valuable.

Emerging Technologies

- **Artificial Intelligence (AI) and Machine Learning:** Understanding the principles and applications of AI and machine learning, which are driving innovations in automation, predictive analytics, and personalization.
- **Blockchain:** Knowledge of blockchain technology and its applications in finance, supply chain management, and secure transactions.
- **Internet of Things (IoT):** Understanding how interconnected devices communicate and operate and the implications for manufacturing, healthcare, and smart cities.

Importance of Digital Literacy and Technology Proficiency

- **Enhanced Productivity:** Proficiency with digital tools and technologies enhances efficiency and productivity, enabling employees to perform tasks more effectively and precisely.
- **Competitive Advantage:** Individuals skilled in the latest technologies have a competitive edge in the job market. Employers seek candidates who can bring innovative solutions and adapt to new technological trends.

- **Adaptability:** As technology evolves, those with strong digital literacy are better equipped to learn and adopt new tools and platforms, ensuring they remain relevant and valuable in their careers.
- **Innovation and Problem-Solving:** A solid understanding of digital tools and emerging technologies empowers individuals to develop creative solutions to complex problems, driving organizational innovation.

STRATEGIES FOR DEVELOPING DIGITAL LITERACY AND TECHNOLOGY PROFICIENCY

Continuous Learning

- **Online Courses and Certifications:** Platforms like Coursera, edX, and Udacity offer courses on various aspects of digital literacy and emerging technologies. Obtaining certifications can demonstrate expertise to potential employers.
- **Workshops and Seminars:** Attending industry workshops and seminars helps individuals stay updated on the latest trends and best practices.

Hands-On Experience

- **Practical Projects:** Engaging in personal or professional projects allows individuals to apply their knowledge and gain practical experience.
- **Internships and Apprenticeships:** Gaining real-world experience through internships or apprenticeships in tech-focused roles provides valuable insights and skills.

Networking and Mentorship

- **Professional Networks:** Joining professional networks and communities, such as LinkedIn groups or industry-specific forums, provides learning opportunities from peers and industry experts.
- **Mentorship Programs:** Seeking guidance from mentors who are experienced in the field can provide personalized advice and support in developing digital skills.

Digital literacy and technology proficiency are critical skills for the future workforce. As technology advances, individuals must develop and continuously update their digital skills to stay competitive and relevant. By focusing on basic digital skills, advanced IT competencies, and knowledge of emerging technologies, individuals can enhance their productivity, drive innovation, and secure a competitive advantage in the evolving job market.

DATA SCIENCE AND ANALYTICS

Introduction

Data science and analytics encompass a set of methodologies, algorithms, and tools used to extract insights and knowledge from data. In today's data-driven world, organizations across various industries rely on data science to make informed decisions, drive innovation, optimize processes, and gain a competitive edge. Data scientists and analysts are pivotal in interpreting complex data sets to uncover patterns, trends, and correlations that inform strategic business decisions.

Key Components of Data Science and Analytics

Data Collection and Integration

- **Data Sources:** Gathering data from various sources, including databases, sensors, social media, and web applications.
- **Data Integration:** Combining and organizing data from disparate sources creates unified datasets suitable for analysis.

Data Cleaning and Preprocessing

- **Data Cleaning:** Identifying and rectifying errors, missing values, and inconsistencies in the data to ensure accuracy and reliability.
- **Data Transformation:** Standardizing and transforming data into a suitable format for analysis, such as normalization or encoding.

Exploratory Data Analysis (EDA)

- **Descriptive Statistics:** Summarizing and visualizing data to gain insights into its distribution, variability, and relationships.
- **Data Visualization:** Creating visual representations (*e.g.*, charts, graphs, heatmaps) to facilitate understanding and interpretation of data patterns.

Statistical Analysis and Modeling

- **Statistical Methods:** Applying statistical techniques (*e.g.*, hypothesis testing, regression analysis) to explore relationships and make predictions based on data.
- **Machine Learning:** Utilizing algorithms and models (*e.g.*, classification, clustering, regression) to analyze data and generate predictions or recommendations.

Data Interpretation and Insights

- **Pattern Recognition:** Identifying meaningful patterns, trends, and anomalies within data sets to extract actionable insights.
- **Predictive Analytics:** Using historical data to forecast trends, behaviors, or outcomes and support decision-making processes.

Data-driven Decision Making

- **Business Intelligence:** Leveraging data insights to inform strategic decisions, optimize operations, and drive business growth.
- **Risk Assessment:** Evaluating risks and uncertainties based on data analysis to mitigate potential challenges and capitalize on opportunities.

Importance of Data Science and Analytics

- **Informed Decision Making:** Data-driven insights enable organizations to make informed decisions based on evidence and analysis rather than intuition.
- **Operational Efficiency:** Optimizing processes and resource allocation through data analytics improves efficiency and cost savings.
- **Competitive Advantage:** Leveraging data to understand market trends, customer preferences, and competitive landscapes gives organizations a competitive edge.
- **Innovation and Product Development:** Identifying opportunities for innovation and new product development based on data-driven insights and market demands.
- **Customer Experience:** Personalizing services and improving customer satisfaction by understanding and responding to customer behaviors and preferences.

STRATEGIES FOR IMPLEMENTING DATA SCIENCE AND ANALYTICS

Data Strategy and Governance

- **Define Objectives:** Clearly define business goals and objectives that align with data science initiatives.
- **Data Quality Assurance:** Implement processes to ensure data accuracy, completeness, and integrity throughout its lifecycle.
- **Data Security:** Establish protocols and measures to protect sensitive information and comply with data privacy regulations.

Infrastructure and Technology

- **Data Infrastructure:** Invest in robust data storage, processing, and retrieval systems (*e.g.*, data warehouses, cloud computing) to handle large-scale data sets.
- **Analytics Tools:** Utilize advanced analytics platforms and tools (*e.g.*, Python, R, SQL, Tableau) to perform complex data analysis and visualization.

Talent Acquisition and Development

- **Skills Development:** Foster a continuous learning and skill development culture among data science teams to keep pace with evolving technologies and methodologies.
- **Interdisciplinary Teams:** Encourage collaboration between data scientists, domain experts, and business stakeholders to leverage diverse perspectives and domain knowledge.

Ethical Considerations

- **Data Privacy:** Ensure compliance with data protection regulations (*e.g.*, GDPR, CCPA) and adopt ethical guidelines for data collection, storage, and usage.
- **Transparency:** Maintain transparency in data-driven processes and decisions to build trust among stakeholders and mitigate risks associated with bias or data misuse.

Continuous Improvement and Evaluation

- **Performance Metrics:** Establish key performance indicators (KPIs) to measure the effectiveness of data science initiatives and their impact on business outcomes.
- **Feedback Loop:** Solicit feedback from stakeholders and end-users to refine analytics strategies, improve data models, and enhance decision-making capabilities.

Data science and analytics are integral to modern organizations seeking to harness the power of data to drive innovation, enhance decision-making, and achieve strategic objectives. By leveraging advanced analytics techniques, machine learning algorithms, and data-driven insights, businesses can gain a competitive advantage, improve operational efficiency, and deliver personalized customer experiences. Implementing a robust data strategy, investing in talent and technology, and adhering to ethical standards are essential for maximizing the benefits of data science and analytics while addressing challenges associated with data complexity and privacy concerns.

ADAPTABILITY AND CONTINUOUS LEARNING

Introduction

Adaptability and continuous learning are critical attributes that enable individuals to thrive in dynamic environments, embrace change, and remain resilient in the face of challenges. These skills are essential in personal and professional contexts, allowing individuals to adjust to new circumstances, acquire new knowledge and skills, and effectively contribute to their teams and organizations.

Key Components of Adaptability and Continuous Learning

Adaptability Skills

- **Flexibility:** Being open and receptive to change and adjusting one's approach or mindset accordingly.
- **Problem-Solving:** Navigating challenges and uncertainties by applying creative and resourceful solutions.
- **Resilience:** Bouncing back from setbacks and maintaining a positive attitude during adversity.
- **Risk Management:** Assessing risks, making informed decisions, and adapting strategies to mitigate potential challenges.

Continuous Learning

- **Curiosity:** Strong desire to explore new ideas, concepts, and experiences.
- **Skill Development:** Actively seeking opportunities to acquire new skills and knowledge relevant to personal and professional growth.
- **Self-Reflection:** Assessing strengths, weaknesses, and areas for improvement to guide learning and development efforts.
- **Adopting New Technologies:** Embracing and mastering new technologies and tools to enhance productivity and effectiveness.

Embracing Change

- **Open-Mindedness:** Being receptive to different perspectives and approaches and willing to consider alternative solutions.
- **Adapting to New Roles:** Transitioning smoothly into new roles or responsibilities and learning quickly on the job.
- **Cultural Awareness:** Understanding and respecting diverse cultural norms, values, and practices in multicultural environments.

Time Management and Prioritization

- **Organization:** Effectively managing time, resources, and priorities to meet deadlines and achieve goals.
- **Adapting Workflows:** Adjusting workflows and processes to accommodate changing priorities and requirements.
- **Balancing Multiple Responsibilities:** Handling multiple tasks and responsibilities efficiently without compromising quality or productivity.

Importance of Adaptability and Continuous Learning

- **Professional Growth:** Continuous learning enhances skills and knowledge, making individuals more adaptable and competitive in the job market.
- **Innovation and Problem-Solving:** Adaptability fosters creativity and innovation by encouraging new perspectives and approaches to challenges.
- **Resilience and Stress Management:** Being adaptable helps individuals manage stress and navigate unexpected changes with confidence and composure.
- **Leadership and Team Dynamics:** Adaptive leaders inspire and motivate teams, fostering a culture of agility and responsiveness within organizations.

STRATEGIES FOR DEVELOPING ADAPTABILITY AND CONTINUOUS LEARNING

Embrace Challenges and Opportunities

- **Seek New Experiences:** Volunteer for challenging projects or assignments that offer opportunities for growth and learning.
- **Step Outside Comfort Zones:** Push boundaries and explore unfamiliar territories to develop resilience and adaptability.

Commit to Lifelong Learning

- **Formal Education:** Pursue courses, certifications, or advanced degrees to acquire new skills and knowledge.
- **Self-Directed Learning:** Engage in reading books, articles, and online resources relevant to professional interests and goals.

Seek Feedback and Reflection

- **Feedback:** Solicit feedback from peers, mentors, or supervisors to identify improvement areas and growth opportunities.
- **Reflection:** Reflect on experiences, successes, and failures to extract lessons learned and apply them to future endeavors.

Network and Learn from Others

- **Mentorship:** Seek guidance and mentorship from experienced professionals who can offer insights and advice based on their experiences.
- **Peer Learning:** Collaborate with colleagues and peers to share knowledge, exchange ideas, and learn from each other's perspectives.

Stay Current and Adapt to Change

- **Stay Informed:** Keep abreast of industry trends, technological advancements, and global developments that impact your field of expertise.
- **Adapt Quickly:** Proactively adjust strategies, plans, and actions in response to changing circumstances or market dynamics.

Adaptability and continuous learning are indispensable skills that empower individuals to thrive in an ever-evolving world. Individuals can enhance their adaptability, resilience, and professional growth by cultivating a mindset of flexibility, embracing change, and committing to ongoing self-improvement. These skills enable individuals to navigate challenges effectively and contribute to innovation, leadership effectiveness, and organizational success. Embracing lifelong learning ensures that individuals remain agile, relevant, and capable of making meaningful contributions to their teams and communities throughout their careers.

INTERPERSONAL AND SOFT SKILLS

Introduction

Interpersonal and soft skills are essential attributes that enable individuals to effectively interact, communicate, and collaborate with others. These skills go beyond technical expertise and play a crucial role in personal development, professional success, and building positive relationships in various contexts, including the workplace, social environments, and personal life.

Key Components of Interpersonal and Soft Skills

Communication Skills

- **Verbal Communication:** Articulating thoughts and ideas clearly and effectively in spoken language.
- **Non-Verbal Communication:** Using body language, gestures, facial expressions, and eye contact to convey messages.
- **Active Listening:** Paying attention to others, understanding their perspectives, and responding thoughtfully.

Emotional Intelligence

- **Self-Awareness:** Understanding one's emotions, strengths, weaknesses, and impact on others.
- **Empathy:** Recognizing and understanding the emotions and perspectives of others and responding with sensitivity.
- **Emotional Regulation:** Managing and controlling emotions to maintain composure and professionalism.

Teamwork and Collaboration

- **Collaborative Problem Solving:** Working effectively with others to identify issues, generate solutions, and achieve common goals.
- **Conflict Resolution:** Resolving disagreements and conflicts constructively while maintaining positive relationships.
- **Supporting Others:** Offering assistance, encouragement, and constructive feedback to team members.

Adaptability and Flexibility

- **Open-Mindedness:** Being receptive to new ideas, perspectives, and changes in circumstances.
- **Adapting to Change:** Adjusting one's approach and behaviors in response to evolving situations and challenges.
- **Resilience:** Bouncing back from setbacks, learning from failures, and maintaining a positive attitude.

Leadership and Influence

- **Decision-Making:** Making informed decisions, considering different viewpoints and implications.
- **Motivation:** Inspiring and motivating others to achieve goals and perform at their best.
- **Influencing Skills:** Persuading and negotiating effectively to gain support and cooperation from others.

Time Management and Organization

- **Prioritization:** Managing tasks and responsibilities based on importance and deadlines.
- **Planning and Goal Setting:** Setting clear objectives, creating action plans, and monitoring progress towards goals.
- **Efficiency:** Using time and resources effectively to maximize productivity and achieve desired outcomes.

Networking and Relationship Building

- **Building Rapport:** Establishing and maintaining positive relationships with colleagues, clients, and stakeholders.
- **Networking Skills:** Developing and nurturing a network of contacts to gain support, advice, and opportunities.
- **Professional Etiquette:** Demonstrating respect, integrity, and professionalism in interactions with others.

Importance of Interpersonal and Soft Skills

- **Effective Communication:** Clear and empathetic communication fosters understanding, trust, and collaboration.
- **Team Effectiveness:** Strong interpersonal skills enhance teamwork, improving productivity and innovation.
- **Conflict Resolution:** Skillful management of conflicts reduces tension and promotes a harmonious work environment.
- **Leadership Development:** Soft skills are crucial for effective leadership, influencing others positively, and driving organizational success.
- **Personal and Professional Growth:** Developing interpersonal skills enhances personal relationships, career advancement, and well-being.

STRATEGIES FOR DEVELOPING INTERPERSONAL AND SOFT SKILLS

Practice and Feedback

- **Role-Playing:** Using simulated scenarios to practice communication, conflict resolution, and negotiation skills.
- **Seeking Feedback:** Soliciting feedback from peers, mentors, or supervisors to identify areas for improvement and refine skills.

Training and Development

- **Workshops and Courses:** Participating in seminars or courses on communication, emotional intelligence, and leadership skills.
- **Self-Study:** Reading books, articles, and resources on interpersonal effectiveness and soft skill development.

Experiential Learning

- **Volunteering:** Participating in volunteer activities to develop teamwork, empathy, and leadership skills while contributing to the community.

- **Cross-Functional Projects:** Taking on projects that involve collaboration with diverse teams to gain practical experience in teamwork and problem-solving.

Networking and Mentoring

- **Networking Events:** Attending professional events and networking opportunities to practice interpersonal skills and expand professional connections.
- **Mentorship Programs:** Seeking guidance and mentorship from experienced professionals to gain insights and perspective on career development and soft skills.

Continuous Reflection and Improvement

- **Self-Reflection:** Reflecting on interpersonal interactions, identifying strengths and areas for growth, and setting personal development goals.
- **Journaling:** Keeping a journal to track progress, document achievements, and reflect on lessons learned from experiences.

Interpersonal and soft skills are foundational to personal and professional success. They enable individuals to communicate effectively, collaborate productively, and navigate relationships with empathy and integrity. By actively developing and honing these skills through practice, training, feedback, and continuous self-improvement, individuals can enhance their effectiveness, leadership capabilities, and overall satisfaction in both professional and personal spheres of life. Cultivating strong interpersonal skills enriches individual interactions, contributes to a positive organizational culture, and fosters long-term career growth and success.

SUSTAINABILITY AND ENVIRONMENTAL AWARENESS

Introduction

Sustainability and environmental awareness are critical concepts in today's global discourse. They focus on responsible stewardship of natural resources and minimizing ecological impacts. As societies grapple with the consequences of climate change, biodiversity loss, and resource depletion, the need for sustainable practices and heightened environmental awareness becomes increasingly urgent. Individuals and organizations play pivotal roles in adopting sustainable lifestyles and business practices to mitigate environmental degradation and promote long-term ecological balance.

Key Components of Sustainability and Environmental Awareness

Environmental Science Fundamentals

- **Ecological Systems:** Understanding ecosystems, biodiversity, and the interrelationships between organisms and their environment.
- **Natural Resource Management:** Knowledge of sustainable practices for conserving and efficiently utilizing water, energy, and raw materials.
- **Climate Change Dynamics:** Awareness of the causes, impacts, and mitigation strategies related to global climate change.

Sustainable Practices

- **Waste Reduction and Recycling:** Implementing strategies to minimize waste generation, promote recycling, and adopt circular economy principles.
- **Energy Efficiency:** Utilizing energy-efficient technologies and practices to reduce energy consumption and carbon emissions.
- **Sustainable Agriculture and Food Systems:** Supporting practices that enhance soil health, conserve water, and minimize chemical inputs in agriculture.

Environmental Policy and Regulation

- **Environmental Laws and Standards:** Understanding local, national, and international regulations to protect the environment and promote sustainability.
- **Compliance and Reporting:** Ensuring adherence to environmental standards and transparent reporting on environmental performance.

Corporate Social Responsibility (CSR)

- **Ethical Sourcing:** Promoting transparency and ethical practices in supply chains to ensure sustainable sourcing of materials and products.
- **Community Engagement:** Engaging with local communities to address environmental concerns and promote sustainable development initiatives.
- **Stakeholder Collaboration:** Collaborating with stakeholders, including governments, NGOs, and communities, to achieve shared sustainability goals.

Impact Assessment and Mitigation

- **Environmental Impact Assessments (EIAs):** Evaluating the potential environmental impacts of projects, policies, or actions before implementation.
- **Risk Management:** Identifying environmental risks and developing mitigation strategies to prevent or minimize adverse impacts.

Importance of Sustainability and Environmental Awareness

- **Resource Conservation:** Promoting sustainable practices ensures the efficient use of resources and reduces waste.
- **Climate Change Mitigation:** Adopting measures to reduce greenhouse gas emissions and combat climate change impacts.
- **Biodiversity Preservation:** Protecting and restoring ecosystems to safeguard biodiversity and ecological balance.
- **Public Health:** Improving air and water quality improves public health outcomes.
- **Long-Term Economic Stability:** Sustainable practices can lead to cost savings, enhance resilience to environmental risks, and improve corporate reputation.

STRATEGIES FOR PROMOTING SUSTAINABILITY AND ENVIRONMENTAL AWARENESS

Education and Awareness Campaigns

- **Public Education:** Raising awareness through campaigns, workshops, and educational programs on environmental issues and sustainable practices.
- **Employee Training:** Providing training and workshops to employees on sustainability initiatives and their role in achieving environmental goals.

Adoption of Sustainable Technologies

- **Renewable Energy:** Investing in renewable energy sources such as solar, wind, and hydroelectric power.
- **Green Building Design:** Constructing and renovating buildings with energy-efficient materials and designs.

Policy Advocacy and Implementation

- **Advocacy:** Supporting policies and regulations that promote sustainability and environmental protection at local, national, and international levels.
- **Internal Policies:** Developing and enforcing internal policies prioritizing sustainability in business operations and decision-making.

Collaboration and Partnerships

- **Industry Collaboration:** Partnering with peers, suppliers, and stakeholders to develop and implement sustainable practices and innovations.
- **Community Engagement:** Engaging with local communities to address environmental concerns and promote sustainable development initiatives.

Continuous Improvement and Monitoring

- **Performance Metrics:** Establishing Key Performance Indicators (KPIs) to measure and track environmental performance and progress towards sustainability goals.
- **Audits and Assessments:** Conduct regular audits and assessments to evaluate sustainability initiatives' effectiveness and identify areas for improvement.

Sustainability and environmental awareness are integral to creating a more resilient and equitable future for all. By embracing sustainable practices, promoting environmental stewardship, and fostering a culture of responsibility, individuals, organizations, and governments can work together to mitigate environmental impacts and ensure the well-being of future generations. Building awareness, implementing effective strategies, and continuously improving practices are essential to achieving global sustainability goals and preserving the planet's natural resources for future generations.

PROJECT MANAGEMENT

Introduction

Project management is the discipline of planning, organizing, and overseeing the completion of specific goals and tasks within a given timeframe and budget. It involves coordinating resources, managing teams, and ensuring project objectives are met efficiently and effectively. Effective project management is crucial for successfully executing projects across various industries, from construction and engineering to information technology and marketing.

Key Components of Project Management

Project Initiation

- **Project Charter:** A document that outlines the project's purpose, objectives, scope, stakeholders, and authorization.
- **Feasibility Study:** An analysis of the project's viability, including technical, economic, legal, and operational considerations.

Project Planning

- **Scope Management:** Defining and documenting the project's objectives, deliverables, and boundaries.
- **Time Management:** Developing a project schedule, including task durations, dependencies, and critical paths.

- **Cost Management:** Estimating project costs and creating a budget to ensure the project remains financially feasible.
- **Resource Management:** Identifying and allocating resources, including team members, equipment, and materials.
- **Risk Management:** Identifying potential risks, assessing their impact, and developing mitigation strategies.
- **Communication Planning:** Establishing a plan for communicating information among stakeholders and team members.
- **Quality Management:** Defining quality standards and processes to ensure project deliverables meet the required criteria.

Project Execution

- **Team Management:** Leading and managing the project team to ensure tasks are completed on time and within scope.
- **Task Coordination:** Assigning and coordinating tasks among team members and ensuring resources are used efficiently.
- **Stakeholder Engagement:** Regularly communicate with stakeholders to keep them informed and involved.

Project Monitoring and Control

- **Performance Tracking:** Monitoring the project progress against the plan using key performance indicators (KPIs) and metrics.
- **Change Management:** Managing changes to the project scope, schedule, or budget in a controlled manner.
- **Issue Resolution:** Identifying and resolving issues that arise during the project to keep it on track.

Project Closure

- **Deliverable Handover:** Completing and handing over the final deliverables to the client or end-users.
- **Project Review:** Conduct a post-project review to assess performance, identify lessons learned, and document best practices.
- **Administrative Closure:** Finalizing all project documentation, contracts, and financial accounts.

Importance of Project Management

- **Efficiency and Effectiveness:** Project management ensures that projects are completed on time, within budget, and to the required quality standards, enhancing efficiency and effectiveness.

- **Risk Management:** Proactive risk identification and mitigation help prevent potential issues from derailing the project.
- **Stakeholder Satisfaction:** Effective communication and stakeholder management ensure that all parties are informed and their expectations are met.
- **Resource Optimization:** Proper planning and allocation of resources prevent wastage and ensure optimal use of available assets.
- **Continuous Improvement:** Post-project reviews and lessons learned constantly improve project management practices and processes.

STRATEGIES FOR DEVELOPING PROJECT MANAGEMENT SKILLS

Formal Education and Certification

- **Project Management Courses:** Enroll in courses and programs that cover project management principles, methodologies, and tools.
- **Certifications:** To validate skills and knowledge, obtain certifications such as PMP (Project Management Professional), CAPM (Certified Associate in Project Management), or PRINCE2 (Projects IN Controlled Environments).

Practical Experience

- **On-the-Job Training:** Gaining experience by working on projects in various roles to understand different aspects of project management.
- **Volunteer Projects:** Participating in volunteer projects to apply project management skills in real-world scenarios.

  Use of Project Management Tools
- **Software Proficiency:** Becoming proficient in project management software such as Microsoft Project, Asana, Trello, or JIRA to plan, track, and manage projects.
- **Technology Integration:** Leveraging digital tools for collaboration, communication, and documentation.

Networking and Mentorship

- **Professional Associations:** Joining project management associations (*e.g.*, PMI, IPMA) to network with professionals and stay updated on industry trends.
- **Mentorship:** Seeking mentorship from experienced project managers to gain insights and advice.

Continuous Learning

- **Workshops and Seminars:** Attending workshops, seminars, and conferences to learn about the latest project management practices and tools.

- **Reading and Research:** Staying informed by reading books, articles, and research papers on project management topics.

Soft Skills Development

- **Leadership:** Developing leadership skills to effectively lead and motivate project teams.
- **Communication:** Enhancing communication skills to ensure clear and practical information exchange among stakeholders and team members.
- **Problem-Solving:** Building problem-solving and critical thinking skills to address challenges and make informed decisions.

Project management is a critical skill that ensures the successful completion of projects across various industries. By mastering the key components of project initiation, planning, execution, monitoring, control, and closure, professionals can enhance their ability to deliver projects effectively. Developing project management skills through education, practical experience, tool proficiency, networking, continuous learning, and soft skills development is essential for achieving project success and advancing in a project management career.

GLOBAL AND CULTURAL AWARENESS

Global and cultural awareness is essential for personal and professional success in today's interconnected world. This awareness involves understanding and appreciating cultural diversity, recognizing global interdependencies, and navigating cross-cultural interactions effectively. As businesses and organizations operate in increasingly multicultural environments, engaging with diverse cultures and perspectives becomes crucial for fostering collaboration, innovation, and mutual respect.

Key Components of Global and Cultural Awareness

Cultural Competence

- **Understanding Cultural Differences:** Recognizing and respecting the customs, traditions, values, and social norms of different cultures.
- **Cultural Sensitivity:** Being aware of and sensitive to cultural differences and avoiding behaviors that might be perceived as disrespectful or inappropriate.
- **Adaptability:** Adjusting behavior and communication styles to accommodate cultural variations in different contexts.

Global Perspective

- **Awareness of Global Issues:** Understanding global challenges such as climate change, poverty, human rights, and economic disparities.
- **International Relations:** Recognizing the impact of political, economic, and social relations between countries and regions.
- **Global Citizenship:** Embracing a sense of responsibility towards the worldwide community and taking action to contribute positively to the world.

Cross-Cultural Communication

- **Effective Communication:** Developing skills to communicate clearly and respectfully with people from diverse cultural backgrounds.
- **Language Skills:** Learning additional languages to facilitate better communication and understanding in multicultural settings.
- **Non-Verbal Communication:** Understanding and appropriately using non-verbal cues such as gestures, eye contact, and body language that vary across cultures.

Cultural Awareness in Business

- **Cultural Intelligence (CQ):** The ability to relate to and work effectively across cultures, including understanding different cultures' business etiquette and practices.
- **Global Market Understanding:** Recognizing and adapting to international markets' cultural preferences and consumer behaviors.
- **Inclusive Leadership:** Leading and managing diverse teams by fostering an inclusive environment that values and leverages cultural differences.

Importance of Global and Cultural Awareness

- **Enhanced Collaboration:** Facilitating better teamwork and collaboration by understanding and valuing diverse perspectives and working styles.
- **Improved Communication:** Reducing misunderstandings and conflicts by effectively communicating across cultural boundaries.
- **Innovation and Creativity:** Leveraging diverse cultural perspectives to drive innovation and creativity in problem-solving and decision-making.
- **Market Competitiveness:** Expanding business opportunities by understanding and responding to the needs and preferences of global markets.
- **Personal Growth:** Enriching personal experiences and broadening worldviews through exposure to diverse cultures and global issues.

STRATEGIES FOR DEVELOPING GLOBAL AND CULTURAL AWARENESS

Education and Training

- **Cultural Studies:** Enrolling in courses or programs focused on cultural anthropology, international relations, or global studies.
- **Workshops and Seminars:** Participating in workshops and seminars on cultural competence, diversity, and inclusion.

Travel and Immersion

- **International Travel:** Traveling to different countries to experience and understand various cultures firsthand.
- **Cultural Immersion Programs:** Participating in cultural exchange programs or living abroad to gain deeper insights into other cultures.

Language Learning

- **Language Courses:** Taking language courses to learn new languages and improve communication with speakers of other languages.
- **Language Practice:** Engaging in language exchange programs or practicing with native speakers.

Diverse Work Environments

- **Multicultural Teams:** Working in or leading multicultural teams to gain practical experience navigating cultural differences.
- **Inclusive Policies:** Advocating for and implementing inclusive policies and practices in the workplace.

Cultural Exchange and Interaction

- **Networking:** Building networks with individuals from diverse cultural backgrounds.
- **Community Engagement:** Participating in community events, festivals, and activities celebrating cultural diversity.

Staying Informed

- **Global News and Media:** Keeping up with global news and media to stay informed about international events and cultural trends.
- **Cultural Literature:** Reading books, articles, and research papers on cultural studies and global issues.

PILOT MODELS REFLECTING EDUCATION 5.0 IN PRACTICE

The evolution from Education 4.0 to **Education 5.0** signals a crucial shift: from tech-centric systems to *human-centric, purpose-driven, and ethically conscious learning*. Education 5.0 aligns with the needs of **Industry 5.0**, integrating values such as sustainability, emotional intelligence, and lifelong learning into pedagogical frameworks (UNESCO, 2021).

Several nations and institutions are actively piloting Education 5.0-inspired models. These examples showcase how innovative, interdisciplinary, and inclusive approaches are taking shape worldwide.

Finland's Phenomenon-Based Learning (PhenoBL)

Finland's education system, led by the Finnish National Agency for Education, pilots **phenomenon-based learning**, where students explore real-world topics like climate change or artificial intelligence across disciplines. This model promotes empathy, systems thinking, and co-creation—key tenets of Education 5.0 (Lonka, 2018). It shifts the focus from rote memorization to holistic understanding, collaboration, and reflective thinking.

Japan's Society 5.0-Inspired Educational Reforms

Japan's government has embedded the *Society 5.0* concept into its education strategy, aiming to harmonize technological innovation with human well-being. Schools are piloting AI-assisted learning, intergenerational knowledge sharing, and personalized digital content, with a strong emphasis on ethical awareness and inclusivity (Cabinet Office, Government of Japan, 2020). These reforms emphasize lifelong learning, human-machine collaboration, and responsible citizenship.

India's National Education Policy (NEP) 2020 Pilots

India's NEP 2020 promotes **multidisciplinary, flexible, and digitally enhanced education** that mirrors Education 5.0 values. Pilot programs such as the *National Digital Education Architecture (NDEAR)* and *Skill Hubs* introduce project-based learning and real-world vocational training from an early age (Ministry of Education, 2020). The policy emphasizes equity, critical thinking, and values education alongside STEM.

University 21's Global Citizenship Program

University 21, a network of global research universities, runs the *Global Citizenship Program* where students engage in digital collaboration across countries to address challenges aligned with the UN Sustainable Development Goals (U21, 2022). These projects develop leadership, cross-cultural awareness, and creative problem-solving skills central to Education 5.0.

Microsoft AI for Good and AI Classroom Series

Microsoft's **AI for Good Challenge** and **AI Classroom Series**, in collaboration with UNESCO and education boards like CBSE India, provide ethical AI training to school students. Participants learn AI design thinking and develop socially impactful projects (Microsoft, 2021). These programs reflect Education 5.0's commitment to technology use for societal good, creativity, and digital equity.

TOWARD SCALABLE EDUCATION 5.0 ECOSYSTEMS

These pilot models illustrate that **Education 5.0 is actionable**, not aspirational. They demonstrate that real transformation emerges when learners are empowered to think critically, act ethically, collaborate across boundaries, and co-create meaningful change. Scaling these practices will require **cross-sector partnerships**, educator training, infrastructure investment, and a renewed commitment to lifelong, values-based education.

CONCLUSION

Global and cultural awareness is vital in the modern world, enabling individuals to navigate and thrive in diverse environments. Individuals can enhance collaboration, innovation, and market competitiveness by developing cultural competence, a global perspective, cross-cultural communication skills, and an understanding of cultural nuances in business. Embracing strategies such as education, travel, language learning, diverse work environments, cultural exchange, and staying informed can significantly contribute to building global and cultural awareness. This skill enriches personal experiences and fosters a more inclusive, respectful, and interconnected world.

REFERENCES

Brynjolfsson, E., McAfee, A. (2014). The second machine age: Work, progress, and Prosperity in a time of brilliant technologies. W.W. Norton & Company.

Cabinet Office, Government of Japan *Society 5.0: Realizing a super-smart society.* Availbale from: https://www8.cao.go.jp/cstp/english/society5_0/

LinkedIn Learning *Future of work: Skills and trends.* Available from: https://www.linkedin.com/learning/future-of-work-skills-and-trends

Lonka, K. (2018). *Phenomenon-based learning: What is it? How does it work?.*Psychology Today Availbale from: https://www.psychologytoday.com/intl/blog/learning-21st-century/201812/phenomenon-based-learning

McKinsey Global Institute *The future of work: Skills outlook for 2030.* Available from: https://www.mckinsey.com/featured-insights/future-of-work/skills-outlook-for-2030

Microsoft *AI for Good: Empowering the next generation.* Availbale from: https://www.microsoft.com/en-in/ai/ai-for-good

Manyika, J., Lund, S., Bughin, J., Robinson, K., Mischke, J., Mahajan, D. (2017). *Jobs lost, jobs gained: Workforce transitions in a time of automation.*McKinsey Global Institute. Available from: https://www.mckinsey.com/featured-insights/future-of-work/jobs-lost-jobs-gained--hat-the-future-of-work-will-mean-for-jobs-skills-and-wages

Ministry of Education, Government of India *National.Educ. Policy.* Availbale from: https://www.education.gov.in/sites/upload_files/mhrd/files/NEP_Final_English_0.pdf

United States Department of Labor. (2022). *Preparing the American workforce for the future: Key trends and challenges in skills development.* Available from: https://www.dol.gov/agencies/eta

U21 (University 21). *Global citizenship program.* Availbale from: https://universitas21.com/global-citizenship

UNESCO *Education for human flourishing: Education 5.0 and the future of learning.* Availbale from: https://unesdoc.unesco.org/ark:/48223/pf0000379338

World Economic Forum *The future of jobs report.* Availbale from: https://www.weforum.org/reports/the-future-of-jobs-report-2022

CHAPTER 3

Key Skills for AI-Era Professionals

Rishikaysh Kaakandikar[1,*] and **Shraddha Vijay Sinkar**[1]

[1] *SaiBalaji International Institute of Management Sciences, Pune, India*

Abstract: The rapid advancement of Artificial Intelligence (AI) is reshaping industries and the workforce, necessitating a paradigm shift in the skill sets required for professionals in the AI era. Key skills for AI-era professionals include a blend of technical proficiency, such as data analysis, machine learning, and coding, alongside strong soft skills like critical thinking, creativity, and emotional intelligence. The increasing integration of AI into business processes demands professionals who can not only operate and innovate within AI-powered environments but also possess a deep understanding of ethics, data privacy, and the human impact of AI. Adaptability, continuous learning, and collaboration are vital for navigating the evolving landscape. By fostering these skills, professionals will be better equipped to thrive in the AI-driven economy, leveraging technology to drive innovation, enhance productivity, and ensure ethical and inclusive growth in organizations and society.

Keywords: Adaptability, AI era professionals, Artificial Intelligence (AI), Critical thinking, Continuous learning, Collaboration, Creativity, Data analysis, Emotional intelligence, Ethics in AI, Innovation, Inclusive growth, Machine learning, Productivity.

INTRODUCTION

Faced with further integration of Artificial Intelligence (AI) into the world of work, people need to acquire technical skills critical to applying artificial intelligence across industries. AI is revolutionizing business processes, automation, and decision-making by having systems mimic human intelligence. Industry professionals must gain basic knowledge of traditional value generation concepts, specific working knowledge of tools and languages used in AI construction, and knowledge of trends in machine learning, data science, and other computational fields. This chapter discusses the technical competencies especially relevant to supporting AI integration: machine learning, data science, coding skills, cloud computing, and deployment methods.

* **Corresponding author Rishikaysh Kaakandikar:** SaiBalaji International Institute of Management Sciences, Pune, India; E-mail krushikesh.123@gmail.com

In the rapidly evolving AI era, professionals must develop a blend of technical, analytical, and soft skills to remain competitive. Technical proficiency, including digital literacy and understanding of artificial intelligence (AI) systems, is paramount as it enables individuals to navigate the complexities of AI-driven environments (Wikipedia, n.d.; arXiv, 2023). Analytical thinking and problem-solving are critical for adapting to AI's impact on industries such as HR, where strategic use of AI reshapes operations (Kaakandikar *et al.*, 2024). Soft skills, such as adaptability, creativity, and effective communication, are equally essential for thriving alongside AI, as they help professionals address dynamic workplace challenges (Etiquette School of New York, n.d.; Creativity at Work, n.d.; EdSurge News, 2024). Leadership skills—including emotional intelligence and the ability to foster collaboration—are increasingly in demand for navigating AI-integrated workflows (Medium, n.d.; Fortune, n.d.; HubSpot, n.d.). Furthermore, professionals must future-proof their careers by cultivating cultural intelligence and diversity-responsive leadership skills to thrive in a globalized AI workforce (Kaakandikar, 2024). As AI continues transforming industries such as finance, healthcare, and marketing, professionals must embrace continuous learning and skill assessment to stay ahead in this transformative era (GoReact, n.d.; Investors.com, n.d.; Business Insider, 2024, July).

UNDERSTANDING AI FUNDAMENTALS

In general, an introduction to the principles of artificial intelligence and many of its subdisciplines forms the basis for AI integration. AI refers to several technologies and approaches that have been designed or are being developed to design systems that can accomplish operations that people can otherwise perform, including reasoning, understanding language, perceiving, and manipulating light or sound.

Key AI subfields include:

Machine Learning (ML)

Although AI is a broader concept that includes all of the advancements studied in this course, ML can be defined as a category of algorithms enabling machines to learn from data. AI ML-based systems allow a system to improve independently without being directly programmed. This involves methods like supervised learning, unsupervised learning, and reinforcement learning.

Deep Learning (DL)

A kind of ML, deep learning utilizes numerous layers in the given neural networks, which imitates the official structure of a human brain for handling large amounts of data, particularly helpful in areas like computer vision, speech visualization, and natural language processing.

Natural Language Processing (NLP)

Natural Language Processing is the ability of a machine to analyze, comprehend, and even interact using natural speech. Some NLP applications include chatbot services, sentiment analysis, language interpretation, and translation.

Computer Vision

This area is in the realm of using signal processing and artificial intelligence to allow machines to understand and come to conclusions using images or video feeds, regularly seen in self-driving cars and facial recognition systems, and used in object recognition.

TECHNICAL COMPETENCY AS A FACTOR FOR AI IMPLEMENTATION.

At the center of AI is machine learning, which allows AI systems to learn from data and inputs. Business professionals who want to implement AI in business solutions must be conversant with the machine learning algorithm and the right time and manner for applying it.

Key skills required to include:

Supervised Learning

This method integrates the utilization of an algorithm labeled using distinct input and output methods. Various working algorithms are used in machine learning, such as linear Regression, Decision Tree, SVM, and KNN.

Unsupervised Learning

However, while dealing with the data, unsupervised learning does not consider labels. The solution has to recognize the patterns or clusters. Several types of clustering include k-means, hierarchical, and Principal Component Analysis (PCA).

Reinforcement Learning

This technique involves training the machines to learn how to choose between options since they 'feel' rewarded whenever an action results in a preferred result. It is commonly used in robotics and game theory.

Data Preprocessing and Feature Engineering

Data preprocessing is a critical step before passing the data to machine learning algorithms. This entails excluding unsupported entries, dealing with extremes, and setting up or standardizing data. Feature engineering directly benefits the accuracy of a created model by choosing the most significant features of a dataset.

Model Evaluation and Optimization

Following model construction in machine learning, it is necessary to assess a model's operational productivity using cross-validation methods, confusion matrices, and accuracy metrics. The general process of model optimization includes a set of techniques, such as hyperparameter tuning, for enhancing the models.

These machine learning techniques are at the core of building the models that integrate AI, and understanding how they work is a crucial factor in AI implementation because failure to apply these techniques correctly causes the implemented solution to provide wrong results.

CLOUD COMPUTING IN THE CASE OF AI SOLUTIONS

Large-scale AI and Machine learning applications pose a computational complexity problem that can be time-consuming and expensive. Cloud computing facilitates AI integration across different platforms since it is a cost-efficient solution for many organizations. The goal is for professionals to know the appropriate cloud computing platforms for AI workloads.

Key cloud platforms for AI to include:

Amazon Web Services (AWS): Services such as SageMaker are available on AWS, which developers can use to create, build, train, and host their Machine Learning models in the cloud. AWS also offers data storage and computing services, which help professionals gain access on tap.

Microsoft Azure: It has tools for solutions like Azure Machine Learning to build AI solutions and Azure Cognitive Services, a collection of AI APIs, including computer vision, language processing, and the rest.

Google Cloud Platform (GCP): Everything from building to deploying machine learning models is made easier through Google Cloud tools like the AI Hub, TensorFlow, AutoML, and many others.

Recognizing cloud platforms' strengths, weaknesses, and capabilities makes AI professionals explore and implement cloud platforms in their solutions, handle big data, and deploy AI models. Real-time collaboration is also achievable with the tools hosted in the cloud since there is no need to install the base of AI applications and services on-site.

To bridge the gap between theoretical skills and practical application, learners must be familiar with current tools in the AI/ML ecosystem. Table **A** summarizes key technical competencies and their associated platforms used across industry and academia (Kelleher, 2019; Chui *et al.*, 2018).

Table A. Summary of future-ready skills and associated tools/platforms.

Skill Category	Specific Skills	Common Tools/Platforms	Application Area
Programming for AI/ML	Python, R, Java	Python, Jupyter Notebooks, Google Colab	Model development, data analysis
Machine Learning	Supervised and unsupervised learning	Scikit-learn, Keras, TensorFlow, PyTorch	Predictive modeling, classification, forecasting
Deep Learning	Neural networks, CNNs, RNNs, LSTMs	TensorFlow, Keras, PyTorch	Image, video, and NLP processing
Natural Language Processing	Text analysis, sentiment analysis, NER	NLTK, spaCy, Hugging Face Transformers	Chatbots, social media analytics
Data Analytics	Data cleaning, EDA, statistics	Pandas, NumPy, Power BI, Tableau	Business intelligence, reporting
Cloud & AI Services	Scalable AI/ML deployment	AWS SageMaker, Microsoft Azure AI, Google AI Platform	Enterprise-scale AI solutions
Computer Vision	Image recognition, object detection	OpenCV, YOLO, TensorFlow	Manufacturing, healthcare, surveillance
Big Data & Data Engineering	ETL, data pipelines	Apache Spark, Hadoop, Kafka	Large-scale data processing
MLOps& Deployment	Model lifecycle, CI/CD, automation	MLflow, DVC, Docker, Kubernetes	Production ML systems
AutoML& Low-Code AI	Automated model selection and tuning	H2O.ai, Google AutoML, Microsoft Lobe	Citizen data science
Soft Skills	Communication, problem-solving, adaptability	Not tool-dependent	Team collaboration, client interaction

This table provides an overview of essential technical skills aligned with AI and machine learning careers, along with widely used tools and platforms supporting their application in professional environments.

TRAINING AI FOR DEPLOYMENT AND MAINTENANCE

Many experts in this field must take additional measures to create an environment for the actual functioning of AI models and maintain them in such surroundings. Successful deployment involves several key processes:

Model Deployment: Experts should know how to distribute machine learning models on the cloud or small and large local servers. This stems from deploying highly trained models into production formats of code and physical incorporation into already existing IT structures.

Model Monitoring: It is also important to point out that after models are deployed, they need to be frequently checked to prevent their performance from deteriorating with time. AI experts must monitor crucial indicators, determine that the models have lost their effectiveness, and update them when required.

Model Updating and Scaling: From the case of AI models, it is evident that such models need to be updated as new data pours in. That's why professionals must master making versions and updating these models without affecting the processes. Moreover, this might require scaling an AI system, another topic involving knowledge of infrastructure or container technologies, such as Docker and Kubernetes.

Automation of Pipelines: To optimize the deployment of AI, it is crucial to automate data and perform data preprocessing, model training, and model deployment. Platforms like Apache Airflow and Kubeflow help orchestrate machine learning pipelines and, thus, help to increase the integration of AI into business processes.

ETHICS AND AI GOVERNANCE

With AI being predominantly incorporated into the economic world and people's lives rapidly, everyone, especially professionals, should grasp its moral consequences. Any AI model that could be developed can constantly adapt the same biases in the data into unfair means. AI practitioners should understand these risks to prevent them from appearing in the first place.

Key ethical considerations in AI integration include:

Bias and Fairness: Promoting AI models with data with no bias and algorithms that benefit both genders.

Privacy: Data protection and confidentiality to protect some of the data used in developing artificial intelligence systems, especially in light of the General Data Protection Regulation (GDPR).

Accountability and Transparency: Creating AI so the user can comprehend the determination process.

It is vitally important that individuals in AI understand ethical issues and apply elements of ethically responsible behavior and best practices for AI that will help define the beneficial interaction of systems with society.

INTEGRATION OF AI AIDS IN THE CLASSROOM: CRITICAL THINKING AND PROBLEM SOLVING

AI and automation technologies have continued to grow at an alarming rate, thus causing significant changes in how industries and people work. Although AI promises enormous opportunities for growth, product differentiation, and better decisions, it also poses challenges that it is well-positioned to address. In this regard, two competencies are paramount: critical thinking and problem-solving when working in environments characterized by AI. AI systems can gather information, analyze it, and draw conclusions. However, human decision-making and judgment are needed to solve sophisticated issues, respond to new circumstances, and develop ethical solutions. Chapter 3 is dedicated to critical thinking and problem-solving skills in using AI to support work: how these skills benefit from AI and how employees can take advantage of AI most effectively.

Defining Critical Thinking for the Age of Artificial Intelligence

Therefore, Critical thinking is the process by which a person comprehends all the information, assesses the available evidence, and arrives at a rational conclusion. In an AI-driven environment, critical thinking involves:

Questioning Assumptions: AI systems are usually built towards specific assumptions, such as the quality of data being fed into the system, the labels used for training, or even the availability of well-defined problem domains. When things turn out otherwise, critical thinkers must be able to challenge the above assumptions and look for other factors or prospects that failed to work out as expected.

Evaluating the Quality of Data: The functionality of AI systems is based on the data they use in learning and making judgments. However, not all data is accurate, comprehensive, and/or uncolored by the biases or agendas of the data gathered.

One of the things that critical thinkers have to do is to evaluate the data being used and be able to note some flaws, distortions, or lack of data.

Assessing AI Output: Thus, an AI system can input the data and make a prediction output. However, it's essential to know that the output produced by an AI system may not be thoroughly perfect or relevant every time. It falls on critical-thinking individuals to assess the result and consider whether an output by the AI is pertinent, precise, or reliable, and if it has shortcomings. For example, this ensures that the developed AI solutions are the right solution to the Problem, hence solving it and providing valuable results.

Ethical Considerations: Moreover, Ethical AI thinking rises as the sources of AI solutions become a part of decision-making. Today's critical thinkers must be able to think about the moral considerations regarding the use of AI applications in various sectors of societal practice, including medicine, banking, law enforcement, and employment services. For example, are the decisions made by an AI model prejudiced? Are there types of biases inherent in the data that influence the results? Is the AI system transparent and available for their usage? These are the kinds of questions critical thinkers need to answer.

Critiques are no longer only about the human decision process but also about how these systems work, their capabilities and non-capabilities, and how they can be put into practice or modified for practice.

Problem-Solving in the Context of the Advance Intelligence-Driven World

Other critical AI skills include Problem-solving. When striving for efficiency and automation, problems can be more complicated, data-based, and intricate than in the past. While AI brings tools that assist in solving these problems, the professionals themselves must apply their critical thinking to frame the issue properly, select the instruments from the AI, and give a suitable meaning to the results.

Here's how problem-solving skills complement AI capabilities:

Defining the Problem Clearly: AI systems are intelligent, but they are good only at solving problems that can have neat and precise solutions. For any problem to be solved, initial analysis divides issues into smaller parts, the first part of problem-solving. Professionals aim to ask the correct questions so they are addressing the proper issues for the given AI system. This involves knowing the Problem and its size, the information that can be gathered, and the solution expected.

Example: In healthcare, an AI system would imply a case where an AI system can diagnose illnesses using patient data. The question in this case is not only "What is diagnosis?" but "What is involved in making the right diagnosis at the right time?" Instead, professionals must present the Problem to match the AI system's capacity.

Selecting the Right AI Tools and Techniques: Regarding problem-solving using artificial intelligence, there are no set recommendations for a proper approach to solving every difficulty. In solving any problem, one has to choose the right tool, model, and technique given the situation that has to be solved. There is no doubt that Machine learning, natural language processing, computer vision, and other AI technologies differ in their application. This knowledge enables one to decide which of the techniques best suits the Problem in question.

Example: When faced with the Problem of identifying customer sentiment, professionals may decide to use NLP algorithms to parse through text data. On the other hand, if the Problem is to find outliers in a given data set, then the machine learning algorithm for outliers may be better suited.

Interpreting AI Solutions and Results: Artificial intelligence systems give solutions rooted in patterns detected in a given dataset. However, the outputs of AI are not always easy to act upon or easily understandable to humans. This is why professionals require problem-solving skills to evaluate the outcome from the models and interpret them for actionable success. This is especially so in areas like finance, where a particular prognosis given by one model over the other has to be put in the context of the whole economy and market.

Example: The machine learning model can answer questions like how likely a customer is to default on the loan, but users, managers, or executives have to make decisions by interpreting these probability values in conjunction with other factors and business criteria, such as customers' records, the state or country's economy, or the company's appetite for risk, for example.

Addressing Bias and Fairness: It is an unfortunate misconception that AI is entirely unimpeachable and free from preconceptions, particularly for the data set provided by history and the history of unfair practices. I want to talk about the problem-solving aspect of AI, which means dealing with any bias within the system to guarantee equal conditions for all.

Navigators of Reason: The critical thinkers and problem-solvers need to determine sources of bias within the data they are analyzing, validate the results for fairness, and develop morally and socially appropriate solutions.

Example: An AI system may be applied to test the candidates during selection. Nevertheless, if the system has been trained on a biased dataset, the result will be inclined toward that particular data type. Among the now-used skills are problem-solving skills that enable professionals to analyse the fairness of the AI system, recognize bias, and resolve it.

Iterative Improvement: There is often not just one Problem that can be solved in a simple and one-for-all manner in AI. In its current state, AI models require updating periodically according to new data, usage feedback, and developing business requirements. Another critical aspect most often ignored when it comes to problem-solving is the aspect of flexibility and fine-tuning of AI solutions.

Such a process is iterative, where adjustments such as improving algorithms or data sets and even retraining of models are conducted to enhance the models' accuracy and performance in the future.

Example: A machine learning recommender system might offer its first recommendation to an e-commerce platform based on little customer information. The more user interactions the system can obtain, the more the need to refine the system for personalizing the results and presenting only relevant suggestions.Automation is defined as using artificial intelligence technologies to develop various versions of repetitive tasks that allow minor or no innovation. Tasks that require little to no creative thinking. In the old days, several companies hired people across different sectors, where they were confined to repeating a plain and straightforward set of activities that called for little or no decision-making, for instance, data entry clerks, report producers, or even call center operators. With innovations in AI, these tasks are automatable, enhancing labor cost control, eliminating human errors, and enhancing productivity. However, while following this line of progression will rid these operations of routine work, it does not render human employees obsolete. It excels in replacing repetitive, mundane tasks that require little to no creative thinking. In the past, employees in various fields were often responsible for performing highly repetitive tasks involving limited decision-making, such as data entry, report generation, or customer support. AI-driven systems can perform these tasks more efficiently, allowing organizations to optimize labor costs, reduce human error, and increase productivity.

However, eliminating routine work does not mean the obsolescence of human employees. However, this automation reduces direct human participation in strategic workplace dealings. Today's Employees are becoming supervisors and recipients of AI-automated outcomes and are responsible for their analysis and enhancement. Human professionals must think in ways that AI systems cannot regarding problem-solving, decision-making, and ethical considerations. Such a shift needs strategic thinking and an understanding of issues such as how to adapt the concept of automation to the firm's outlook on strategic objectives rather than simply applying automation.

Memorization tasks vs strategic thinking: With routine and well-defined work being eased by technologies, the core of problem-solving remains a realm for the worker. Ane functions that require little to no creative thinking. In the past, employees in various fields were often responsible for performing highly repetitive tasks involving limited decision-making, such as data entry, report generation, or customer support. AI-driven systems can perform these tasks more efficiently, allowing organizations to optimize labor costs, reduce human error, and increase productivity.

While automation takes over repetitive, structured tasks, strategic problem-solving remains a domain where human workers excel. Any advanced AI system is limited because it will not be as good at handling uncertain, less well-defined, or new situations where more breadth of analysis, synthesis, and creative judgment may be required. Strategic planning is the intricate exercise of reading into patterns and dependencies and gaining the capability to develop a solution that meets organizational requirements in the short and long run. Active thinking. In the past, employees in various fields were often responsible for performing highly repetitive tasks involving limited decision-making, such as data entry, report generation, or customer support. AI-driven systems can perform these tasks more efficiently, allowing organizations to optimize labor costs, reduce human error, and increase productivity.

THE ROLE OF DECISION-MAKING ABILITIES IN AUTOMATION

Even though automation opens the possibility to perform numerous operations at very high speed, it does not possess the context humans do. The contextual awareness that humans bring to the table. Unlike AI systems, people can learn about events and trends within their field and assess their impact on the business when making decisions; the system, however, cannot understand these changes. Therefore, strategic decision-making has continued to be a critical domain where people bring value quickly and efficiently; it often lacks the contextual awareness that humans bring to the table. AI systems can analyze data, identify patterns, and suggest actions based on historical information. However, they lack a broader understanding of the underlying context, such as shifts in market conditions, customer behavior, or regulatory changes.

As such, strategic decision-making remains a crucial area where humans contribute value. Hence, the professional implications of AI in the future workplace will require such human professionals to interpret these outputs in light of extant problems and decision-making. This includes deciding if an AI suggestion is relevant, calculating risks for an AI suggestion, and considering the ethical ramifications of working with robots.

For instance, in a marketing department, AI tools may automate customer segmentation, content personalization, and ad targeting. While AI systems can generate valuable insights, marketers must use strategic thinking to interpret those insights within the broader context of their brand, customer preferences, and market conditions. They must decide how to leverage AI recommendations to create personalized, engaging campaigns that resonate with customers while maintaining the brand's ethical standards and values.

Moreover, in industries like automotive, where autonomous driving technology is rapidly developing, strategic decision-making involves ensuring the system is technically sound, adheres to safety regulations, minimizes risks, and contributes to creating more innovative, safer transportation systems. Technical professionals at every level, from engineering to design, are also faced with considerations of safety, compliance, and social acceptance in the development of new AI technologies, concerns that the AI system does not simply perform beyond the scope of the mandate.

CUSTOMER SERVICE: AI CHATBOTS AND HUMAN EMPATHY

AI-powered chatbots and virtual assistants are increasingly used in customer service to handle routine inquiries, process orders, and provide basic troubleshooting. These systems are well-suited for quickly and effectively answering most of the different questions customers might have. But customer service is not only about answering questions; it is about empathizing and feeling with the customer that AI does not have.

Case Example: AI Chatbots in customer support: Sephora, H&M. For simple tasks like tracking orders, product inquiries, and other general questions about the company or product, AI chatbots are used. Such bots are generated from massive customer data and can effectively solve issues without involving human beings. For instance, a customer might ask the organization about their order status; the chatbot can pull up the details and inform the client about their order status.

Nonetheless, when customers make complex decisions or become upset owing to some product issue, for instance, a damaged product or a problem with a bill, AI chatbots may fail to address customers' needs—analyzing vast amounts of financial data, identifying trends, and predicting stock prices, market movements, or creditworthiness. While AI is an effective tool for data analysis, human insight is crucial when interpreting these predictions and applying them to real-world scenarios.

Despite the AI's ability to detect anomalies, human oversight is necessary to assess whether the flagged transaction is fraudulent or a legitimate purchase.

AI might flag a transaction as suspicious because it is unusual, but the financial analyst might recognize that the customer is traveling or shopping for a unique item. This contextual judgment is essential to reducing false positives and providing a better customer experience.

Furthermore, AI in finance is used to predict market trends and inform investment decisions. However, market predictions are often influenced by external factors—such as geopolitical events, market sentiment, and government policies—that AI models may not fully comprehend. Human financial analysts use their expertise to factor in these unpredictable elements, adjusting AI recommendations accordingly. This collaboration between AI's efficiency and human insight enables better decision-making in volatile markets.

MARKETING

Artificial Intelligence for Personalized Experience and Human Imagination: AI in marketing entails creating, targeting, analyzing, and optimizing advertisement marketing, targeting consumer behavior, and optimizing campaigns. With the help of artificial intelligence, customer information may be analyzed quickly, and customers' buying habits may also be easily recognized, so that organizations can target content and advertisements based on customer preferences. However, creativity and strategic conduct remain powerful for creating engaging marketing engagements that appeal to the clientele.

Case Example: AI in Content Personalization: There are some ways in which companies like Amazon and movie or music streaming providers like Netflix leverage AI. With the help of AI, customers' activity on the website and their previous purchases, as well as ratings, are used to filter the most suitable products or content. This personalization raises the number of clients who engage with a product and the likelihood of their purchase because of customized suggestions that may be more timely and relevant. But still, they are incapable of providing the creativity required to design an interesting marketing strategy or an advertisement. Lyze consumer behavior and optimize campaigns.

AI systems can quickly process customer data and identify patterns in purchasing behavior, allowing companies to deliver highly personalized content and advertisements. However, creativity and strategic thinking are essential for developing compelling marketing campaigns that resonate with consumers.

SUPPLY CHAIN MANAGEMENT: AI FOR OPTIMIZATION AND HUMAN OVERSIGHT

AI optimizes supply chains by forecasting demand, predicting delivery times, and managing inventory. These systems rely on data collected in the past and current inputs to suggest the best approaches to optimizing resource utilization. AI assists in the decision-making process; however, human intervention is required to address disruptions, manage teams, and make decisions involving ethical issues such as labor relations and environmental concerns.

Case Example: AI in Logistics and Human Cooperation Organizations like Amazon and DHL use AI in their logistics department. Computational intelligence powers forecasts, demands changes, selects the best delivery routes, and monitors inventory to prevent stockouts or superfluous storage. This assists in keeping business costs low and customers happy, as would have been seen in the case of Apple Maps.

However, complications that the AI program can't cover may happen in certain conditions, like emergencies (landscape changes, for example, natural disasters, strikes, or political instabilities). Customer data and identify patterns in purchasing behavior, allowing companies to deliver highly personalized content and advertisements. However, creativity and strategic thinking are essential for developing compelling marketing campaigns that resonate with consumers.

Case Example: AI in Content Personalization Retailers like Amazon and streaming services like Netflix use AI to personalize the user experience. AI algorithms analyze customer behavior, such as browsing history, past purchases, and ratings, to recommend products or content tailored to individual preferences. This personalization increases customer engagement and conversion rates by delivering more relevant and timely recommendations.

However, AI alone cannot generate the creativity to craft captivating marketing campaigns or advertisements. Human marketers must apply strategic thinking to integrate AI-driven insights into more extensive creative campaigns. While AI can provide data on which products are likely to sell well, it is up to marketers to design compelling narratives, create attractive visuals, and devise campaigns that emotionally resonate with the target audience.

For instance, AI may help identify which products are trending. However, human insight is needed to create a campaign that compellingly highlights those products, taking into account brand messaging, market conditions, and cultural trends. Creative professionals must also evaluate the ethical considerations of using AI for personalized marketing, ensuring that consumer data is handled responsibly and with consent.

COMMUNICATION SKILLS

The last one of their studies aims to bridge the gap between AI and humans by developing intelligent interfaces that adapt to the users. With the ever-increasing use of AI in organizations, there is a need to ensure proper communication between humans and the system. The need for clear communication between humans and machines is growing. Although AI can be handy as these systems can analyze vast amounts of data and propose the results, they cannot present this information in an actionable and comprehensible format for all the parties that need it. This is where human communication skills come in— especially when you have to translate just what the AI has calculated to other coworkers, clients, or managers who may not understand all the algorithms involved.

Qualified people responsible for transmitting the outcomes and possibilities of AI solutions will be vital, as the need for clear communication between humans and machines grows. While AI systems can process data and offer insights, they cannot communicate these findings in a manner that is easily understandable to all stakeholders. This is where human communication skills come in—particularly in explaining complex AI-generated insights to non-technical colleagues, customers, or decision-makers.

Professionals who can effectively communicate the results and potential of AI technologies will be indispensable. For instance, data scientists and AI specialists must be able to present the findings of the AI system in ways understandable to other people, specifically those in management roles, to clarify where the AI is operating and where the AI is coming up short, as well as how the insights from the AI system are valuable. Because of ambiguity, consumer value may not be well understood, and decisions concerning AI may be made incorrectly. Further, communication is essential in many customer-serving roles because virtually all activities customers encounter – from initial contacts to complaints and refund requests – may involve AI systems. Systems can process data and offer insights, but they cannot communicate these findings in a manner that is easily understandable to all stakeholders. This is where human communication skills come in—particularly in explaining complex AI-generated insights to non-technical colleagues, customers, or decision-makers.

INTRODUCTION TO COLLABORATION AND TEAMWORK IN AN ADVANCED TECHNOLOGY-ENHANCED CONTEXT

There are possibilities that AI and automation may transform teams. However, they cannot overcome the importance of Cooperation. Learning communication between humans and machines grows. While AI systems can process data and offer insights, they cannot communicate these findings in a manner that is easily understandable to all stakeholders. This is where human communication skills come in—particularly in explaining complex AI-generated insights to non-technical colleagues, customers, or decision-makers.

Effective communication becomes even more critical in customer-facing roles as AI systems are used for everything from initial customer inquiries to troubleshooting and follow-ups. While AI can provide standard responses, human agents are needed for situations requiring personalized explanations or clarity. Human communication is essential for building relationships, trust, and understanding, whether explaining a complex technical process, calming an upset customer, or persuading a client.

WORKING WITH EMOTIONAL INTELLIGENCE (EQ)

Insights for Achieving Success in the Age of Artificial Intelligence: Emotional intelligence (EI) is an individual difference construct that involves knowledge of and regulation of one's own emotions and the feelings of others. Ability to understand and influence the emotions of others. Each of these components of emotional intelligence covers empathy, self-awareness, the ability to manage emotions, and skills, which are characteristically optimal in the current world dominated by artificial intelligence applications. There's nothing wrong with an algorithm's ability to analyze data and make appropriate conclusions while ignoring emotions. As a result, EQ is of increasingly heightened importance in managing human-worker relations while wearing the hat of AI systems.

One of the most vital components of EQ is Empathy, which is critical in customer-related fields such as customer service and healthcare, as well as leadership. The ability to recognize, understand, and manage one's own emotions, as well as the ability to understand and influence the feelings of others. EQ includes key components such as empathy, self-awareness, emotional regulation, and social skills, which are increasingly important in the AI-driven workplace. While AI can process data and make decisions based on logic, it lacks the emotional understanding and awareness that humans bring to interactions. Therefore, EQ is becoming even more valuable to bridge the gap between human employees and AI systems.

Empathy in the Workplace

Empathy, a cornerstone of EQ, is critical in customer service, healthcare, and leadership. While AI might do an excellent job in operations or customer services, answering FAQs, or handling orders, it cannot address multifaceted, contentious, racially/sexually charged issues. For example, AI may help identify a particular disease in the medical industry. However, the healthcare social worker or doctor may be the one to embrace a patient experiencing the news. Likewise, in a customer service context, while much of the first-level interaction could be computer-driven, when a customer is angry and needs to be pacified, feelings make that happen.

Over time, as cognitively-enabled intelligent entities gradually increase their application across significant organisations, human beings with high EQ will be the ones to nurture sensitive forms of customer relations, solve intricate and individualised problems with appropriate comforting words that are hard to come by with AI the emotions of others. EQ includes key components such as empathy, self-awareness, emotional regulation, and social skills, which are increasingly important in the AI-driven workplace. While AI can process data and make decisions based on logic, it lacks the emotional understanding and awareness humans bring to interactions. Therefore, EQ is becoming even more valuable to bridge the gap between human employees and AI systems.

Self-awareness and Emotional Regulation

Self-awareness, another key element of EQ, involves recognizing and understanding one's emotions and how they influence thoughts and behaviors. In the workplace, self-awareness allows individuals to manage stress, stay focused, and make clear decisions, even amid uncertainty or technological change. As AI systems are integrated into various industries, the speed of change can be overwhelming for many workers. Professionals with high emotional intelligence will be better equipped to adapt to these changes, managing their emotions and guiding their teams through the transition.

Emotional regulation is equally essential as AI and automation introduce new dynamics in the workplace. Professionals must manage their emotions as they work with increasingly sophisticated AI systems. In leadership roles, emotional regulation is critical for maintaining a calm, focused approach to problem-solving and decision-making, especially in high-pressure situations. Emotionally intelligent leaders can set an example for their teams, helping others navigate the complexities of working alongside AI.

The Importance of Teamwork in an AI-Driven World. AI can enhance individual productivity, but teamwork is central to organizational success. While AI helps complete one or the other task, collaboration, including interdisciplinary, involves communication that cannot be solved without the actual interaction of people. Interpersonal dexterity becomes critical in manufacturing disciplines where interaction with different functions is required. Professionals must use AI tools proactively; at the same time, they must also be able to come up with their own shared experience, common sense, and innovative thinking.

Effectiveness of AI and Human combined teams: New, organized AI working in the organizational environment does not exclude Cooperation nor eliminate the need for collaboration. However, it alters the maximized form of teamwork, as known before. When AI is applied to tasks within a process, another team gains time to perform challenging, imagination-demanding, and critical functions that should not be outsourced to an MT. The team members are now working with both people and AI systems; hence, they must know how to apply an AI tool while being aware of its effect and impact from an ethical perspective.

For example, in an engineering team within an organization engaged in new product development, AI may support the computation of data, modeling, and forecasting of new product design. As AI takes over routine tasks, human teams can focus on more complex, creative, and strategic work that requires human input. Team members must now collaborate with humans and AI systems, ensuring they understand how to use AI tools effectively while providing human insights, judgment, and ethical considerations.

For instance, in an engineering team working on a new product design, AI can assist with data analysis, simulations, and product performance predictions. However, machine intelligence alone cannot turn the results into a new design that would address the market demand; this requires human intervention. Moreover, in care teams, AI can identify possible courses of treatment based on all patients' information. Still, doctors, nurses, and other healthcare workers try to achieve the best patient results.

Competent teams must also be able to change and embrace new propositions. Integration into workflows does not eliminate the need for collaboration. Instead, it changes the nature of teamwork. As AI takes over routine tasks, human teams can focus on more complex, creative, and strategic work that requires human input. Team members must now collaborate with humans and AI systems, ensuring they understand how to use AI tools effectively while providing human insights, judgment, and ethical considerations.

Interdisciplinary Cooperation as AI advances in sophistication, it will penetrate the organizational structure, affecting marketing and finance, product development, and operations departments. As AI takes over routine tasks, human teams can focus on more complex, creative, and strategic work that requires human input. Team members must now collaborate with humans and AI systems, ensuring they understand how to use AI tools effectively while providing human insights, judgment, and ethical considerations.

Cross-Disciplinary Collaboration

As AI becomes more sophisticated, it will be used across various organizational departments, from marketing and finance to product development and operations. Experts from multiple fields need to come together and incorporate the findings of AI to make their decisions. For instance, while marketing AI can predict consumer behavior to potentiate marketing strategies, the strategy will have to be worked on with data scientists and customer service personnel. Multi-disciplinary work will also create a need for enhanced communication processes as practitioners from different fields share information and have a sorted aim and objective. As AI takes over routine tasks, human teams can focus on more complex, creative, and strategic work that requires human input. Team members must now collaborate with humans and AI systems, ensuring they understand how to use AI tools effectively while providing human insights, judgment, and ethical considerations.

ETHICAL LEADERSHIP IN THE AGE OF AI

AI and automation raise essential ethical questions, such as how AI is trained, the potential for algorithm bias, and how data privacy is protected. Organizational and functional heads must oversee AI in an acceptable way to the organization and society. Firstly, ethical leadership in the context of AI means knowing the technical possibilities of AI and its potential consequences within an organization and the possible consequences for employees, customers, or even the whole society.

For instance, in the employment area of recruitment, AI is shared as a tool that provides recommendations for candidates based on CV analysis. It is primarily responsible for managing people, making decisions, and guiding organizations through various challenges. Today, AI systems can assist with data analysis, decision-making, and task automation. However, leadership remains crucial in navigating the complexities of working with AI and ensuring that AI is used ethically and responsibly.

INSPIRING AND MOTIVATING TEAMS

In the face of rapid technological change, employees need leaders who inspire confidence and motivate them to embrace new technologies. In the AI age, one has to be capable of painting a picture, getting people to sign up for that picture, and making sure that everyone understands why AI is essential for creating that picture.

As discussed below, leaders should also know how to give support during change processes. Leadership as well. Previously, leaders were primarily responsible for managing people, making decisions, and guiding organizations through various challenges. Today, AI systems can assist with data analysis, decision-making, and task automation. However, leadership remains crucial in navigating the complexities of working with AI and ensuring that AI is used ethically and responsibly.

High-fidelity simulation of different healthcare situations: In healthcare, AI can accurately predict a wide range of scenarios based on an individual's medical history, genetics, and lifestyle.cal history, genetic data, and lifestyle factors. They allow AI to create individual programs for the treatment, and a doctor gets a list of options that can be used. In other words, AI can propose one or more medications or one or multiple treatment protocols depending on previous cases and study results.

These now contribute to medical studies and drug development using artificial intelligence-operated simulations. These simulate potential outcomes based on a patient's medical history, genetic data, and lifestyle factors. By running simulations, AI can generate personalized treatment plans, offering doctors a range of options. For example, AI can suggest various drug combinations or treatment regimens based on previous patient data and clinical research.

These AI-driven simulations also assist in medical research and drug discovery. They use AI to predict how new drug compounds will behave within biological systems, thus making it easier to discover disease treatments. By automating the simulation process, researchers can explore more possibilities than by doing it manually, hence coming up with more out-of-the-box solutions.

Using AI to Improve Decision-Making: There is often an element of decision-making based on available information in many creative problem-solving areas. Potential outcomes are based on a patient's medical history, genetic data, and lifestyle factors. By running simulations, AI can generate personalized treatment plans, offering doctors a range of options. For example, AI can suggest various drug combinations or treatment regimens based on previous patient data and clinical research.

These AI-driven simulations also assist in medical research and drug discovery. AI can predict how new drug compounds will interact with biological systems, speeding up the process of finding new disease treatments. By automating the simulation process, researchers can explore more possibilities, leading to more innovative solutions.

ENHANCING DECISION-MAKING WITH AI-GENERATED INSIGHTS

Creative problem-solving involves making complex decisions based on available data in many fields. AI outperforms humans in pattern recognition and computation of big data, thus improving resolution arrival and encouraging improved creativity solutions.

a. Focusing on Business Strategy and Market Analysis, AI can easily model Market trends, consumer behaviors, and competitive forces, generating insightful information to help businesses develop distinctive, creative strategies. By running simulations, AI can generate personalized treatment plans, offering doctors a range of options. For example, AI can suggest various drug combinations or treatment regimens based on previous patient data and clinical research.

These AI-driven simulations also assist in medical research and drug discovery. AI can predict how new drug compounds will interact with biological systems, speeding up the process of finding new disease treatments. By automating the simulation process, researchers can explore a more significant number of possibilities, leading to more innovative solutions.

Creativity and problem-solving through the lens of the future of artificial intelligence•. That said, there is more to be seen from AI shortly, with more efficient creativity in problem-solving. Problem-solving will become even more pronounced. Next-gen AI systems will be more intuitive in reality with humans and more accurate in their thought generation and development.

AI is not the idea killer, but it can help to refine the idea generation process. Future AI systems will better understand human emotions, creativity, and intuition, allowing them to generate even more sophisticated and personalized ideas. Human creativity and AI's computational power will enable professionals to solve problems more innovatively, efficiently, and effectively.

AI is not a replacement for human ingenuity but a tool that can enhance the creative process. AI's ability to evaluate and analyze data, create concepts, model solutions, and ultimately offer insights gives professionals from various industries a new dimension of creativity

Refocusing on Building and Developing Entrepreneurial Capabilities during the Age of Disruption Brought about by Artificial Intelligence. The Artificial Intelligence (AI) Age has become the new reality for industries, changing their models and approaches and offering fresh angles for idea genesis and development. Systems will better understand human emotions, creativity, and intuition, allowing them to generate even more sophisticated and personalised ideas. Human creativity and AI's computational power will enable professionals to solve problems more innovatively, efficiently, and fully.

The rise of Artificial Intelligence (AI) is transforming industries, reshaping business models, and introducing new ways of thinking, creating, and innovating. This paper considers AI and how it has impacted traditional businesses to define the modern meaning of entrepreneurship. Contemporary entrepreneurs face the challenges of a dynamically evolving digital environment and must learn how to manage AI to create sustainable value.

In the disrupted world of AI, becoming an entrepreneur means adopting a perspective that is flexible, innovative, analytical, and, most importantly, savvy about using AI assets strategically. Human emotions, creativity, and intuition allow them to generate even more sophisticated and personalised ideas. Human creativity and AI's computational power will enable professionals to solve problems more innovatively, efficiently, and fully.

DEVELOPING CREATIVITY AND CREATIVITY SKILLS

Long ago, the focus was made on how creative people are, at times, made to be enterprising; however, in the future, it will be more important still. In the age of AI, it has become even more critical. AI does not come as a substitute for creativity because creativity is the cornerstone of any entrepreneurship process; AI is an assistant that allows an entrepreneur to focus on creativity while providing data, insight, and other information within a few seconds.

New creative application for Artificial Intelligence: AI should be seen by entrepreneurs as a tool that strengthens human imagination. It is the art of entrepreneurship, and in the age of AI, it has become even more critical. AI does not replace the need for creative thought; instead, it empowers entrepreneurs to be more creative by enabling them to experiment, analyze data, and derive insights at unprecedented speeds.

HARNESSING AI FOR CREATIVE SOLUTIONS

Entrepreneurs should embrace AI as a powerful tool that can augment human creativity. It can replace repetitive work, which may be uninteresting for an entrepreneur, and save time invested in productive work. In product development, for instance, it enables the examination of customer preferences for products, trends, and market deficiencies that assist entrepreneurs in developing products and services.

AI can also benefit creative business people in designing, marketing, and content production. In the age of AI, readership becomes even more critical. AI does not replace the need for creative thought; instead, it empowers entrepreneurs to be more creative by enabling them to experiment, analyse data, and derive insights at unprecedented speeds.

The construction and management of partnerships in AI are the realms of individual invention and Cooperation. Creating collaborative ecosystems. Today's entrepreneurs need to develop webs of relationships for AI solutions, talents, and tools. AI is complex and may take one to engage IT, art, and business professionals. Working in tandem with other nationally and internationally recognised AI experts. Businesspeople should seek alliances with AI developers, data scientists, and technology firms to provide them with the knowledge required to incorporate those technologies properly. Of partnerships that enable them to access AI expertise, resources, and technologies. AI is complex, and leveraging it often requires interdisciplinary collaboration between technologists, designers, marketers, and other experts.

COLLABORATING WITH AI EXPERTS

Entrepreneurs should consider forming partnerships with AI developers, data scientists, and tech companies to gain the expertise needed to implement AI solutions effectively. This collaboration helps entrepreneurs understand how AI can be applied to their specific business needs and how to integrate AI into their existing workflows.

Entrepreneurs can join AI-focused communities, attend industry conferences, and collaborate with academic institutions to stay updated on the latest developments. Thus, AI ecosystem relations provide early-stage entrepreneurs with access to the necessary resources, novel research, and inside information for the subsequent progress of their businesses.

ETHICAL CONSIDERATIONS AND RESPONSIBLE AI USE

Ethical considerations are increasingly important as AI becomes more integrated into business operations. Businesspeople should understand AI's risks and moral problems, including data protection, prejudice, and responsibility. The Lack of

Advanced Organizational Ethical AI Practices: Self-employed people should adopt ethical uses of Artificial Intelligence that encourage fairness, transparency, and accountability issues. Collaborative ecosystems. Entrepreneurs today must build networks of partnerships that enable them to access AI expertise, resources, and technologies. AI is complex, and leveraging it often requires interdisciplinary collaboration between technologists, designers, marketers, and other experts.

By working with AI experts, entrepreneurs can identify their industry's most effective tools and platforms and avoid common pitfalls in AI adoption. For example, AI-powered tools like ChatGPT or TensorFlow require technical expertise to be used effectively. Entrepreneurs can partner with AI professionals or invest in upskilling their teams to leverage these technologies.

LEVERAGING AI COMMUNITIES AND NETWORKS

Entrepreneurs can join AI-focused communities, attend industry conferences, and collaborate with academic institutions to stay updated on the latest developments. By building relationships within the AI ecosystem, entrepreneurs gain access to valuable resources.

CONCLUSION

The future of work demands a blend of technical proficiency, continuous learning, and strong interpersonal skills. The rapid pace of technological change requires individuals to be adaptable and committed to lifelong learning. By focusing on in-demand skills such as digital literacy, critical thinking, and collaboration, individuals can position themselves for success in the dynamic workplace of tomorrow. Organizations must also foster a culture of innovation and inclusivity to support their workforce. Together, these strategies ensure resilience and growth in an evolving global economy.

REFERENCES

ArXiv. (2023). Artificial intelligence impact on the labour force: Searching for the analytical skills of future software engineers. Available from: https://arxiv.org/abs/2302.13229.

Business Insider. (2024, July). These are the top skills you need to help land a job in AI. Available from: https://www.businessinsider.com/ai-top-skills-you-need-get-job-2024-7.

Business Insider. (2024, October). Leaders from BCG, Infosys, AARP, and more predict how AI will transform top companies in the next 12 months. Available from: https://www.businessinsider.com/leaders-discuss-ai-technology-transform-company-workflows-unlock-employee-potential-2024-10.

Bundle Skills Blog. (n.d.). 10 skills you need to thrive in the AI era (and they're not what you think!) Available from: https://bundleskills.com/blog/10-skills-you-need-to-thrive-in-the-ai-era-and-th-yre-not-what-you-think.

Creativity at Work. (n.d.). Future-proofing your workforce: Essential skills for the AI era. Available from: https://www.creativityatwork.com/the-must-have-skills-you-need-to-compete-with-a-i-and-automation/.

Chui, M., Manyika, J., & Miremadi, M. (2018). Notes from the AI frontier: Insights from hundreds of use cases. McKinsey Global Institute. Available from: https://www.mckinsey.com.

Etiquette School of New York. (n.d.). 10 soft skills essential for thriving in the AI era. Available from: https://etiquette-ny.com/soft-skills-for-the-ai-era/.

EdSurge News. (2024, January 24). The critical power skills needed for the AI era. Available from: https://www.edsurge.com/news/2024-01-24-the-critical-power-skills-needed-for-the-ai-era.

Fortune. (n.d.). 4 key skills for the AI era. Available from: https://brand-studio.fortune.com/servicenow/--key-skills-for-the-AI-era/.

GoReact. (n.d.). Ultimate guide to skill assessment in the AI era. Available from: https://get.goreact.com/resources/ultimate-guide-to-skill-assessment-in-the-ai-era/.

HubSpot. (n.d.). 11 skills you need in the AI era [Data + Tips]. Available from: https://blog.hubspot.com/marketing/skills-needed-in-ai-era.

Investors.com. (n.d.). Learning the ABCs of AI can bolster your career. Available from: https://www.investors.com/news/management/leaders-and-success/learn-ai-can-bolster-your-career/.

Kaakandikar, R., & Rangade, A. (2019, November 9). A study on job satisfaction of employees in an organization. Think India Journal - VicharNyas Foundation. Available from: https://thinkindiaquarterly.org/index.php/think-india/article/view/10355.

Kaakandikar, D.R. (2020). *Impact of artificial intelligence on our society. Zenodo.* [http://dx.doi.org/10.5281/zenodo.13683725]

Kaakandikar, D.R. (2020). *Study of product branding with digital marketing. Zenodo.* [http://dx.doi.org/10.5281/zenodo.13683782]

Kaakandikar, R., Gawande, R.P., Deshmukh, V.A., Raskar, S., Mulani, H.I. (2024). The strategic significance of artificial intelligence (AI) in HR operations and management. *European Economic Letters, 14*(3), 1424-1433.
[http://dx.doi.org/10.52783/eel.v14i3.1907]

Kelleher, J.D. (2019). *Deep Learning..* MIT Press.
[http://dx.doi.org/10.7551/mitpress/11171.001.0001]

Kaakandikar, D.R. (2024). Cultural intelligence pedagogy in management education: Nurturing diversity-responsive leaders. *Zenodo.*
[http://dx.doi.org/10.5281/zenodo.13705855]

Kaakandikar, R., Kaushik, K., Tiwari, P., Ningule, S.S. (2024). *Fintech and blockchains trends in the financial sector..* Bentham Science Publishers.
[http://dx.doi.org/10.2174/97898152568331240101]

Medium. (n.d.). 10 essential leadership skills in the AI era. Available from: https://sertiscorp.medium.com/10-essential-leadership-skills-in-the-ai-era-b94e0adf94cd.

PhonakPro Audiology Blog. (n.d.). What new skills do HCPs need to be 'future-ready' in the era of AI? Available from: https://audiologyblog.phonakpro.com/what-new-skills-do-hcps-need-to-be-futur--ready-in-the-era-of-ai/.

Reddit. (2024). Needed career advice in 2024 in the era of AI. Available from: https://www.reddit.com/r/artificial/comments/18ydqwi/needed_career_advice_in_2024_in_the_era_of_ai/.

The Australian. (n.d.). AI gains to soar after 'second innings' of use: Capgemini. Available from: https://www.theaustralian.com.au/business/ai-gains-to-soar-in-the-second-innings-of-use-capg-mini/news-story/e3bb2fdd1221f7294b6944a2a634b1fb.

The Times. (n.d.). How to prepare students for a digital-first world. Available from: https://www.thetimes.co.uk/article/how-to-prepare-students-for-a-digital-first-world-5xg079crc.

The Wall Street Journal. (n.d.). IT unemployment hits 6% amid overall U.S. jobs growth. Available from: https://www.wsj.com/articles/it-unemployment-hits-6-amid-overall-u-s-jobs-growth-bc2f2915.

Tiwari, P., Kaakandikar, R., Bhosale, S.S., Nirmala, K., Kasar, B. (2024). A critical study of behavioural factors affecting mutual funds investors with special reference to Pune District. *Economic Sciences, 20*(2), 47-61.
[http://dx.doi.org/10.69889/667gf640]

Wikipedia. (n.d.). Digital literacy. Available from: https://en.wikipedia.org/wiki/Digital_literacy.

CHAPTER 4

Soft Skills Reengineering with Education 5.0: Hype and Hope

Madhavi Udaybhan Shamkuwar[1,*], Jayesh Katkar[1] and Ajinkya Sanjay Bhalerao[2]

[1] *Zeal Institute of Business Administration, Computer Application & Research, Pune, India*

[2] *Sinhgad Institute of Management and Computer Application, Pune, India*

Abstract: 50% of all employees will need reskilling by 2025, as adoption of technology increases, according to the World Economic Forum's Future of Jobs Report. The World Economic Forum has identified the top 10 skills for 2025. The first six on the list are soft skills crucial for job candidates: analytical thinking and innovation, active learning and learning strategies, complex problem-solving, critical thinking and analysis, creativity and originality, and leadership and social influence.

According to the United Nations Sustainable Development Goals (SDGs), Goal 4 focuses on inclusive and equitable quality education, along with the promotion of lifelong learning opportunities for all. It also eliminates gender disparities in the education sector and ensures equal access to all levels of education for vulnerable persons. The need for soft skills is at different levels for kindergarten, high school, raduates, and post-graduates. High School students need soft skills like Communication, Flexibility and adaptability, Dependability, Problem-solving and critical thinking, Teamwork, and collaboration. On the other hand, the University Graduate and Post Graduates need different soft skills such as Emotional Intelligence and Relationship Building, Adaptability and Resilience, Leadership and Team Management, Ethical Decision-Making, and Social Responsibility.

Soft skills Statistics, According to the Nasscom 2022 report, Soft skills such as communication, collaboration, and critical thinking are highly sought by IT industry. But many candidates lack these Soft skills: communication, collaboration, and critical thinking, due to which they are unemployable.

Keywords: Critical thinking, Emotional intelligence, Inclusive education, Leadership, Reskilling, Soft skills, Sustainable employment.

* **Corresponding author Madhavi Udaybhan Shamkuwar:** Zeal Institute of Business Administration, Computer Application & Research, Pune, India; E-mail: madhavi.sh@gmail.com

Rishikaysh Kaakandikar, Keshav Kaushik, Supriya Pathakm Madhavi Shamkuwar & Vivek Mishra (Eds.)

INTRODUCTION

Soft skills play a very important role in the 21st century, as they are crucial in almost all avenues of life, from schools and colleges to universities, and even in landing a job and retaining it, having good soft skills. Along with hard skills, all individuals are expected to have good soft skills. Soft skills can be attained from formal and informal education and different jobs require different soft skills but most job descriptions require many similar and common soft skills from fresher graduate, post graduate and experienced candidates alike such as good communication skills in written as well as oral form, Time Management, Emotional intelligence, Collaboration, Adaptability, Active listening etc to name a few as per various national and international Employability skills survey reports. Many jobs in BPO, FMCG, Education, and the Teaching sector mainly require a very good command of soft skills along with a few hard skills[Burbekova, 2021.

Education 1.0 to Education 5.0 has been tremendous, but still more emphasis needs to be given to soft skills, as their importance is utmost today. In Education 5.0, the main emphasis is given to soft skills development such as creativity, critical thinking, communication, and collaboration as these soft skills are essential for Society 5.0 (Vieira *et al.* 2023) (Tavares *et al*, 2023). One of the main root causes of unemployment is the soft skills. With the introduction of Education 5.0, this gap may be reduced, and the candidate may be able to fulfil the soft skill expectations of Recruiters and the Corporate World alike.

Placement statisticsfrom an perspective of On campus placement and Off campus placement the existing data, literature available tells us that many Graduate and Post Graduate are not able secure a On campus placement and also struggle to get Off campus placement because they not only lack knowledge of their core field or specialization but they also lack soft skills that form an essential part of job description and are very valuable to MNC Multinational Companies and Corporate World Alike. In 2024, IIT Bombay released an exit survey report stating that 6.1% of the students are unplaced. According to the annual Unstop Talent Report 2024 (unstop, 2024), Only 7% of Indian colleges achieve full campus placement also 91% student feel college curricula offer an adequate level of preparedness for a job, but, 66 per cent of recruiters and 42 per cent of university partners feel the skill gap and lack of preparation, respectively, are major challenges in campus recruitment (freepressjournal,2024) (bizzbuzz.news, 2024) [5, 6].

The broad sections of the Chapter involve a literature review, both theoretical and pictorial visual literature review; Global perspective, Local perspective, Education 5.0 and its impact on Soft Skills, Case studies, and a Conclusion.

Objectives:

1. To study the importance of Soft Skill Re-engineering in today's context with respect to Education 5.0
2. To Study the Impact of Soft Skills on Campus Placement.

Systematic Literature Review

Systematic literature review. is a thorough and methodical technique to examine the body of literature on a particular subject or research issue. It is a systematic, logical, and structured process of searching, selecting, evaluating, and creating relevant research in order to represent a body of knowledge suited to objectives. The quantitative method process involves a literature search, analysis and review with meta-analysis for the content. (Kraus *et al*, 2020) (Haddow, 2018) PRISMA (Preferred Reporting Items for Systematic Reviews and Meta-Analyses) is a universally designed set of guiding principles Galvão *et al*., 2015) to conduct systematic reviews and meta-analyses.

The PRISMA protocol consists of a diagrammatic research flow diagram and a set of checklists to have a logical and sequential flow to conduct systematic reviews and meta-analyses so that the data is visualized for better understanding and providing transparency to the stakeholders (Epidemiol, 2025). The PRISMA flow diagram is a four-phase pictorial representation of a process—identification, screening, eligibility, and inclusion of research articles (Sina, 2022) [11].

Selection Process for the Study

An extensive process to collect the relevant research papers was carried out, where the research papers were searched from IEEE xplore(ieeexplore.com) [12]. The main purpose was to determine the scope of the research regarding different education standards, particularly Education 5.0 development. Second, the criteria were extended to articles published in the Soft Skills, as the study focused on the combined research articles towards unifying Education and Soft Skills. Lastly, this process was also restricted to a 6-year research development to make the process more authentic and updated; the last PRISMA checklist was used for the current study. For this purpose, some important keywords were used in the search process, like 'Education 1.0, 2.0, 3.0, 4.0, 5.0, Industry 1.0, 2.0, 3.0, 5 (Google trends) [13]. The result thus generated is depicted in Fig. (**1**). The x-axis depicts the year of search, and the y-axis depicts the search count of three keywords: Education 5.0, Industry 5.0, and Society 5.0.

Visual Literature Review

A visual literature review was conducted using various research tools, which create an image for the review at a glance. Some of the research tools used included Carrot2, Open Knowledge Maps, and Citation Gecko.

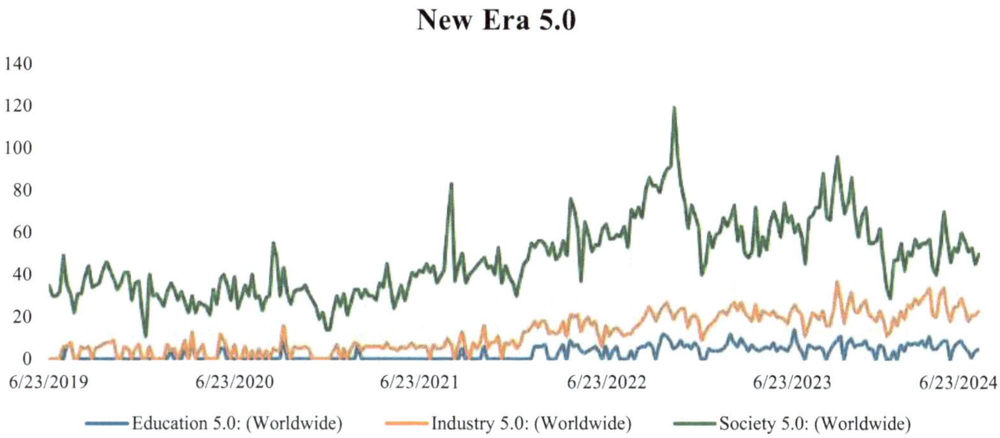

Fig. (1). Google Trends comparison of Education 5.0, Industry 5.0, and Society 5.0.

Carrot2 is an open-source engine that creates clusters of documents and represents them pictorially in the form of a pie chart and a tree map chart (Carrot2.org) [14]. The search involved filter options such as language, country, and sources. It supports five different languages, namely, English, French, German, Italian, and Spanish, and supports documents from eight countries, i.e., Austria, France, Germany, Great Britain, Liechtenstein, Italy, Spain, and Switzerland. The visual representation (Figs. (**2** and **3**)), consist of a pie chart and treemap, it was generated using the keywords "Education 5.0" and "Soft skills." The result highlights critical thinking, problem-solving, employability performance, and the importance of learning, among other key concepts.

Citation Gecko is a graphical user interface (citationgecko.com) [15] exclusively created to find the association between various papers for the related keyword(s). A visual representation was generated when using the search keyword "smart agriculture," and all the papers were considered to be "seed papers." As a result, two visualizations were created: a list of seed papers and clusters of seed papers and related papers. The visualizations are given below:

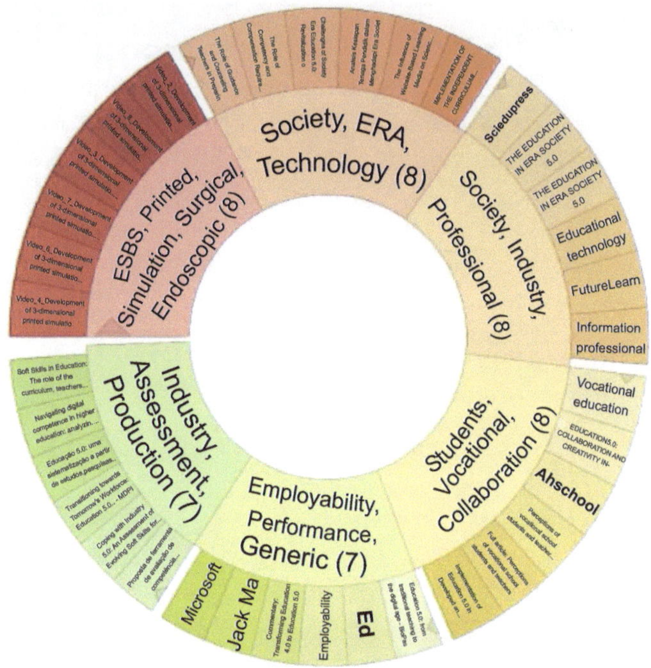

Fig. (2). Piechart generated with the keyword "Education 5.0 and soft skills" from Carrot2.

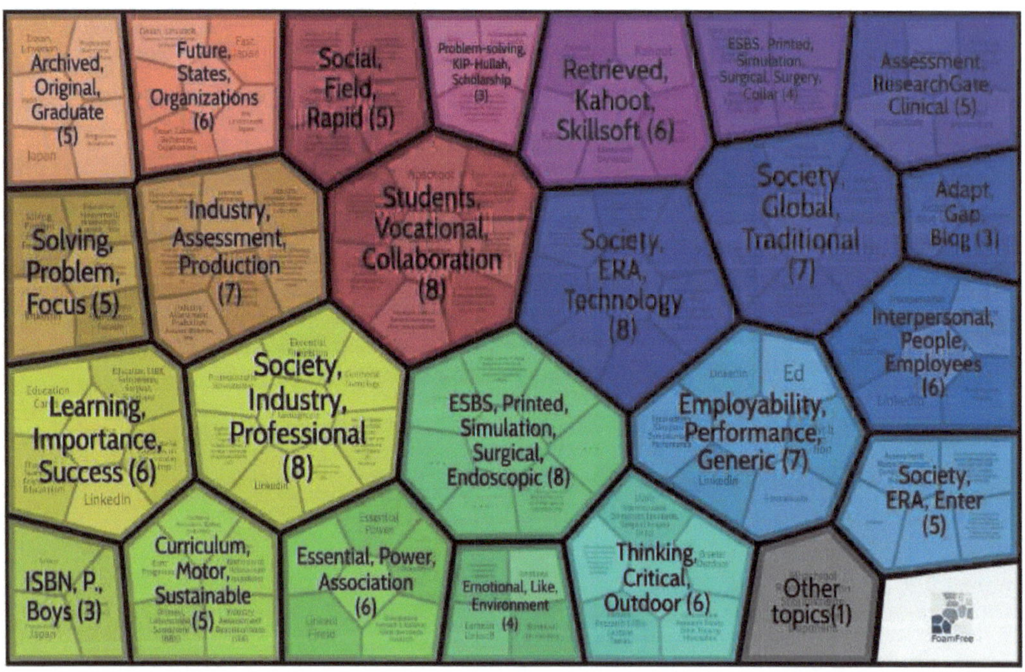

Fig. (3). Tree map generated with the keyword "Education 5.0 and soft skills" from Carrot2.

A visual representation was generated when using the search keyword "Education 5.0 and soft skills," and all the papers were considered to be "seed papers." As a result, two visualizations were created: a list of seed papers and clusters of seed papers and related papers in Figs. (**4** and **5**). Fig. (**4**) depicts a list of seed papers generated with the keyword "Education 5.0 and Soft skills" from Citation Gecko. Fig. (**5**) depicts a cluster of seed papers generated with the keyword "Education 5.0 and Soft skills" from Citation Gecko. The x-axis shows the number of research papers, and the y-axis represents the year of publication.

The visualizations are given below:

Bibliometric Analysis

Bibliometric analysis is a type of quantitative data analysis used to evaluate the impact and visibility of research articles within a given subject or discipline (Aria *et al*, 2017) (Waltman, 2019). (It analyses the impact and visibility of research articles within a particular subject or domain, as shown in the figure.

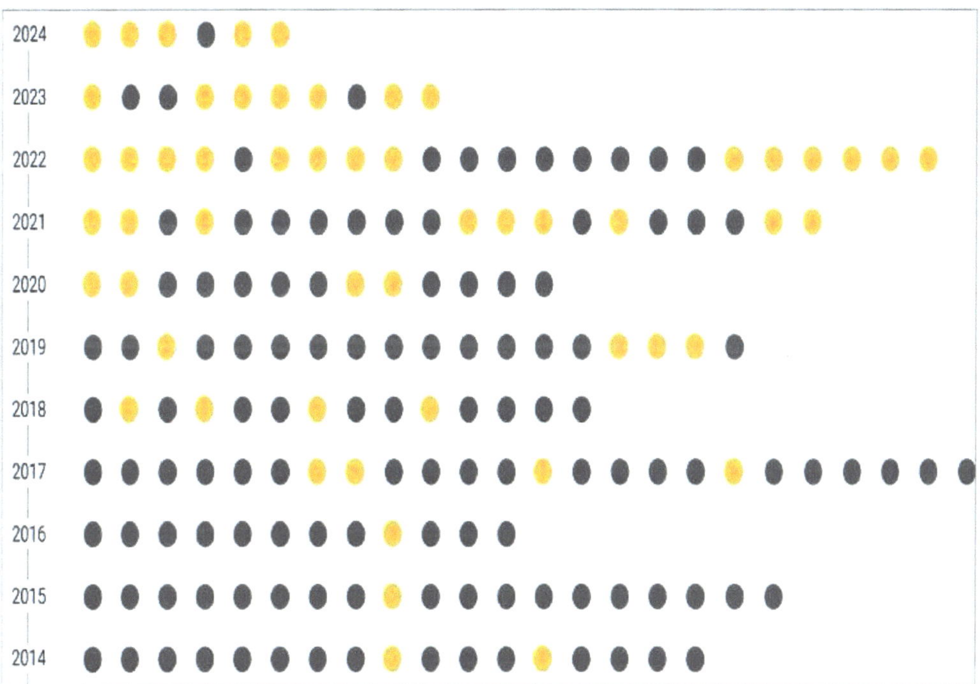

Fig. (4). List of seed papers generated with the keyword "Education 5.0 and Soft skills" from Citation Gecko.

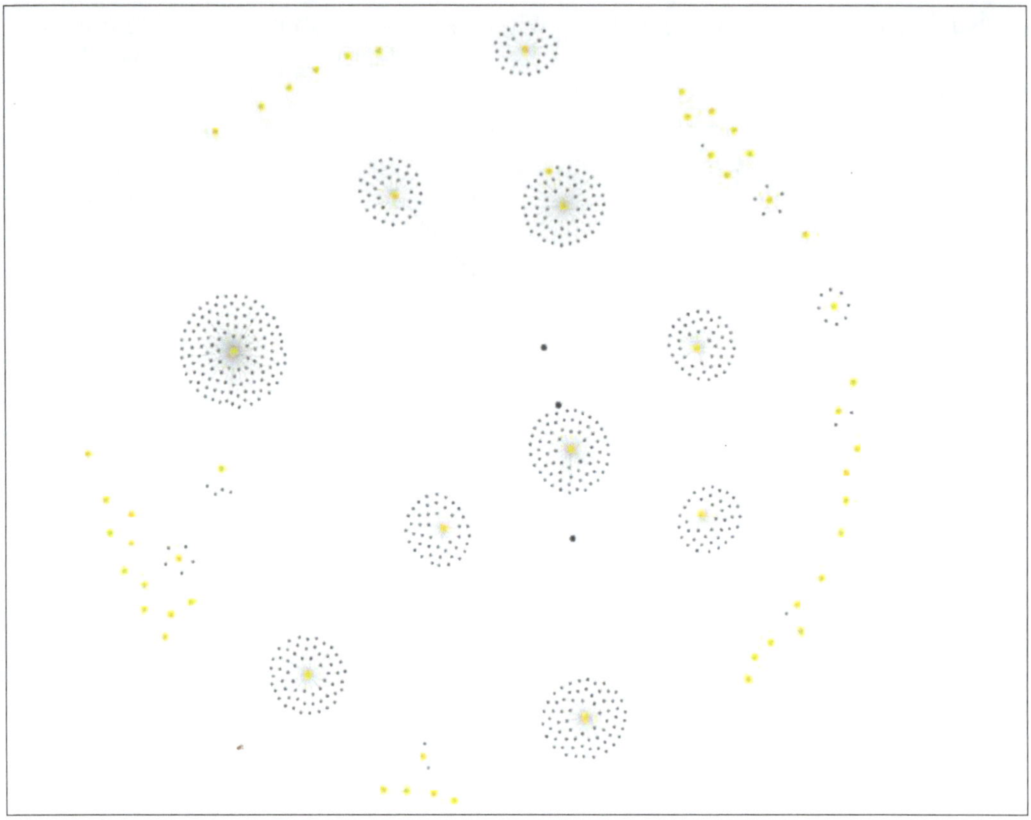

Fig. (5). Cluster of seed papers generated with the keyword "Education 5.0 and Soft skills" from Citation Gecko.

Database

The first step towards a bibliometric study is to determine a suitable research database source for the study and then extract the relevant set of documents from it. The sources can be single or multiple. IEEE Xplore (Institute of Electrical and Electronics Engineers) is a rich research documents database providing research articles in multiple domains in the field of electrical engineering, electronics, computer science, and related domains (Sinha, 2021) [18]. It consists of more than six million documents comprised of research articles, conference proceedings, standards, books (Cobo, 2015), and other publications. IEEE Xplore considers data in various formats: .ris, .bib, and .csv. Using IEEE Xplore basic and advanced search to apply the foundation queries or filter methods, respectively. Moreover, the platform offers the option to filter the research articles on the basis of multiple dimensions, including article type, publication year, author, affiliation, publication title, publisher, conference location, standard status, standard type, and publication topics. This enables users to search for terms.

The study for bibliometric analysis was performed utilizing IEEE Xplore, a research database providing a collection of various research articles. The keywords "Education 4.0" and "Soft skills" were provided in the advanced search filtering technique of IEEE Xplore.

Data from about 2,000 research articles have been selected and downloaded in .csv format. The search strategy is one of the challenges in bibliometric studies; it involves constructing a 'valid search query'with an aim to balance the quality, relevance, and significant number of retrieved documents and eliminating the 'false positives', i.e., irrelevant documents are not retrieved in the search. This challenge is particularly relevant in the field of Education 5.0 and Soft skills, where Education 5.0 is the recent education phenomenon, and soft skills are known from the time of origin.

Advance Search 20896

When searching the databases using the criteria ("Full text only": Education 5.0) OR ("Document Title": Soft Skills), the results yielded very few relevant articles. This scarcity is likely because Education 5.0 is a recent top limited documents were being generated for the research study, this signifies the exclusivity of the research topic. Due to this limitation, the range of years is not specified; a total of 1000 research articles, as the strong foundation of the literature review, were exported into Microsoft Excel. The data that were exported contained information on the yearly increases in publications, document types, languages, nations, authors, institutions, journals, citations, and funding organizations. In order to construct network visualization maps, the acquired literature was also exported to the VoSviewer.The VoSviewer derived its name from the Visualization Of Similarity (VOS) (www.vosviewer.com) algorithm that visually represents these networks among concepts, terms, and keywords. The distances between pairs of objects on the map accurately depict their similarity with high mathematical precision (Tang, 2002) [21]. We utilize a feature that enables the creation of co-occurrence networks using keywords, as shown in Fig. (**6**). These networks have the nodes represented by the keywords as specified by the authors, and the links between nodes are proportional to the degree of co-occurrence between the associated words (Yong, 2002). To showcase the strength of international research collaboration, we utilized the Total Link Strength (TLS) metric, which is automatically generated by VOS viewer when mapping the research activities of selected nations. The TLS metric serves as an indicator of the extent and intensity of collaborative ties between countries in the research domain, providing valuable insights into the level of cooperation and knowledge exchange among nations. The degree of international research collaboration is inversely correlated with the TLS, with a higher TLS value indicating more collaboration. The common authors of the research articles were known, and a network among them was established, as given in Fig. (7). This network is called 'Author analysis.' During that period, the primary research focused on the advancement of digital agriculture, encompassing several key areas. These included the establishment of comprehensive databases, the formulation of metadata standards, the innovation of monitoring systems, the design of decision-making forecasting systems, and the implementation of information dissemination systems. These research endeavors aimed to lay the foundation for the progress and development of digital agriculture (Hirsch, 2005) (Oswalt, 2020) . However, in the subsequent years, there was a scarcity of publications focusing on these subjects. For the study, the quality journals were considered with high citations and h-index (Yadav, 2020) [25]. The top 10 active bibliometric indicators were displayed. Tableau was utilized to depict the annual growth of publications. In the present study, the authors devised an extensive and inclusive search query with the aim of retrieving all relevant documents pertaining to different aspects of agriculture.

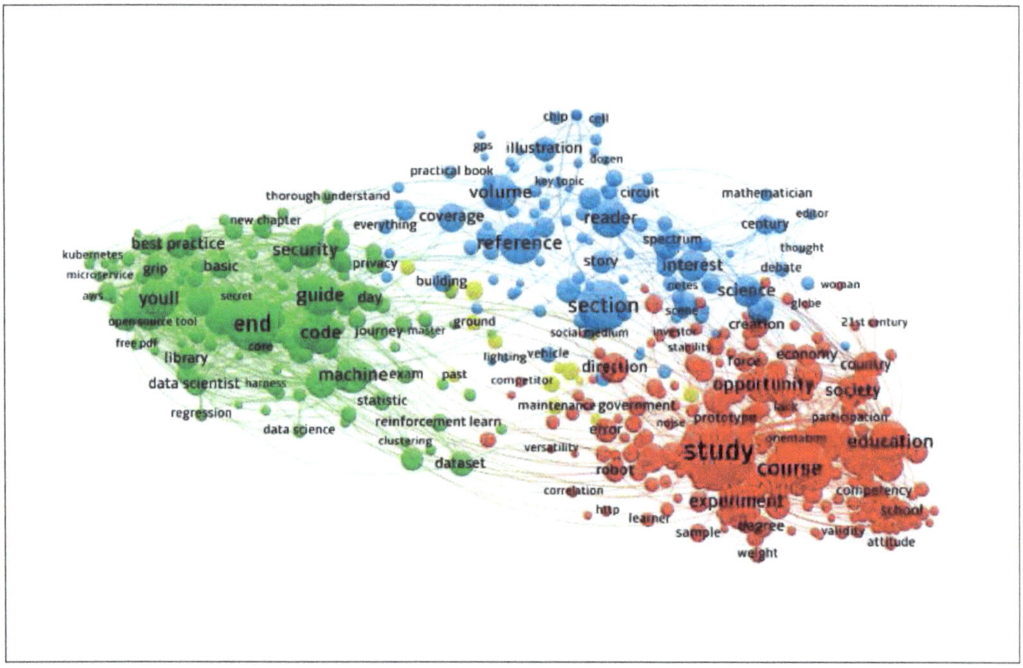

Fig. (6). Co-citation analysis

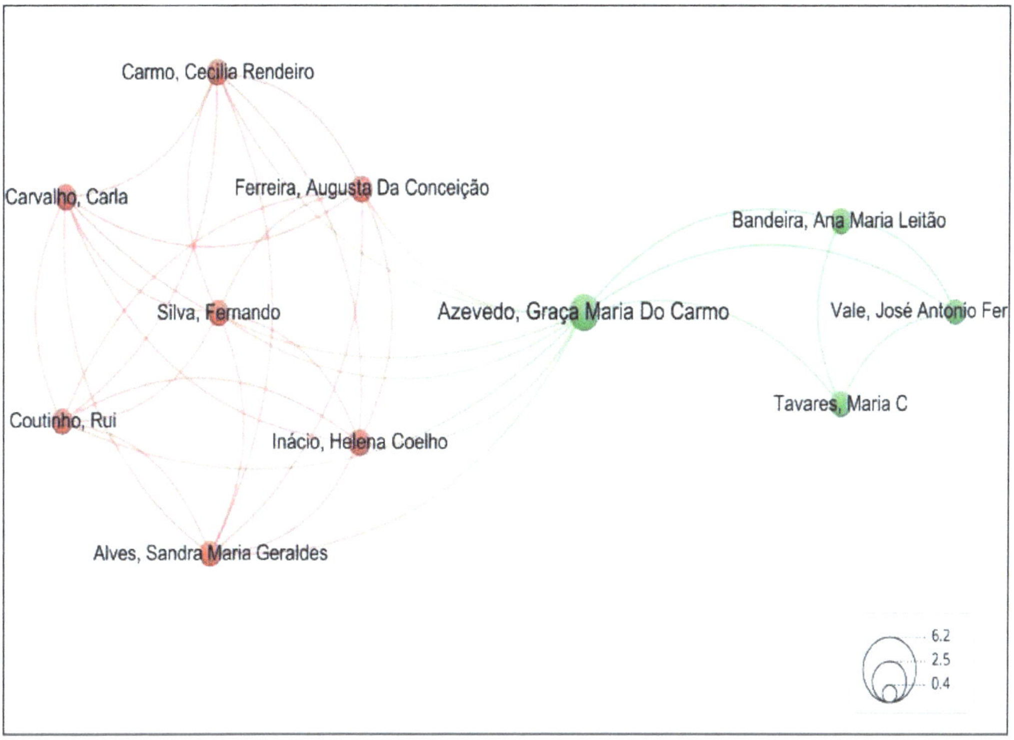

Fig. (7). Author analysis

As given in Fig. (**6**), the 4 clusters of keywords were formed: red, green, blue, and yellow, with 243, 142, 119, and 15 keywords. The diameter of the bubbles in the co-citation analysis is directly proportional to the no. of occurrences of keywords. The cluster description is shown in Table **1** and a pie-chart in Fig. (**8**). As shown in Fig. (**8**), the red cluster has 47% of the total keywords.

Table 1. Clusterwise research items.

Cluster	Color	Items
Cluster1	Red	243
Cluster2	Green	142
Cluster3	Blue	119
Cluster4	Yellow	15
Total:		**519**

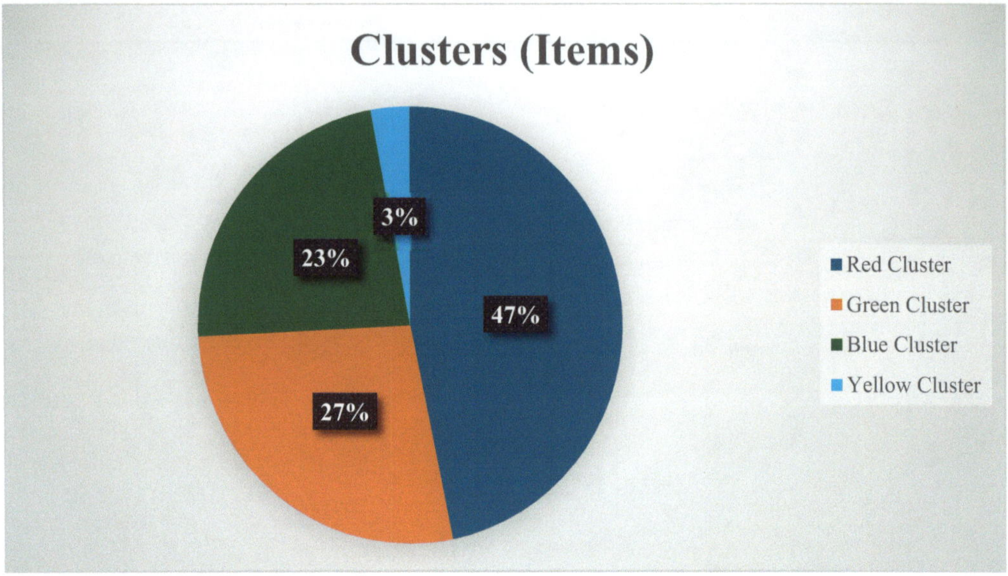

Fig. (8). Pie chart of various clusters.

GLOBAL PERSPECTIVES:

Adoption of Education 5.0 in Various Nations:

Early adoption of technologies in Education is necessary to improve the overall education system of a nation, which in turn expedites the soft skill development. A combination of hard skills and soft skills needs to be taught to all students to prepare them for the corporate world. In the Global Context, Saudi Arabia, Malaysia, and China have started the adoption of new technologies in their education systems. Countries like Zimbabwe and Sri Lanka are still lagging behind in terms of Education 5.0 Integration and implementation.

Saudi Arabia and Malaysia are selected from the developed countries in the region, where Education 5.0 is being implemented successfully. Zimbabwe and Sri Lanka are selected from the developing countries to observe and assess the deployment gap of Education 5.0. A study has found that there is a big need to revisit the current situation of the implementation of Education 5.0 in developing countries. The study could be further extended to investigate, based on quantitative and qualitative methodologies in different areas, as pointed out in the study, to observe and assess some other latent concerns that are affecting the implementation of Education 5.0 in developing countries (Alharbi, 2023).

The shift to Education 5.0 is necessary for Enhancing Higher education in India, as globally many Countries, developed and developing alike, are integrating

Education 5.0 into the Curriculum of their Schools, colleges, and universities to stay ahead of other Countries that are still struggling with integrating Education 3.0 and Education 4.0. In India, the education sector is going through a rapid overhaul as subjects related to Innovation and entrepreneurship, Global competitiveness are now being taught all over the nation. The recent NEP National Education Policy in India aims to bring uniformity in the Indian Education System.

As per the PwC ASSOCHAM Report July 2024 (ASSOCHAM Report, 2024), India's education sector is undergoing a massive transformation as seen by recent events such as the refined National Education Policy (NEP), an increase in global industry connections with educational institutes, and the use of new-age technologies. This marks a point in time where our shared histories will head towards the point where education becomes even more powerful, personalized, and freely accessible than it used to be.

Education 5.0 and its impact on Soft Skills-Case Studies:

In Zimbabwe, there is a skill gap between the level of skills of students and the level of skills demanded by the corporate world. So recently, the Ministry of Higher Education has decided to integrate Education 5.0 into the current Curriculum of Higher Education. However, many challenges exist in implementing Education 5.0 into the current Curriculum (SimbarasheMunikwa *et al.*, 2022). A communicative approach should be embraced in the Education 5.0 curriculum revision process. The guiding team needs to be supported through internal staff development workshops and to effectively engage in in-depth dialogue with academics for effective sharing of ideas on Education 5.0 Curriculum revision. Academics need to be capacitated in curriculum development processes. The academics may be encouraged to enroll for the Post Graduate Diploma in Higher Education to support their Professional development. The Education 5.0 curriculum revision process needs to be allocated reasonable timelines (Nair *et al* 2009).

CONCLUSION

Many studies carried out on recent pass-out graduate candidates indicate that they do not possess the minimum required hard skills and soft skills needed to attain a job in the Industry. This skill gap is large, especially for fresh graduates. So the need for re-engineering arises for fresh graduates with respect to their hard skills and soft skills.

Research on Engineering Students shows that student learning outcomes indicate that university graduates do not possess important skills required by employers/industry, such as communication, decision-making, problem-solving, leadership, emotional intelligence, social ethics skills, as well as the ability to work with people of different backgrounds.

REFERENCES

Alharbi, A.M. (2023). Implementation of Education 5.0 in developed and developing countries: A comparative study. *Creat. Educ., 14*(5), 914-942. [http://dx.doi.org/10.4236/ce.2023.145059]

Aria, M., Cuccurullo, C. (2017). bibliometrix : An R-tool for comprehensive science mapping analysis. *J. Informetrics, 11*(4), 959-975. [http://dx.doi.org/10.1016/j.joi.2017.08.007]

Burbekova, S. (2021). Soft skills as the most in-demand skills of future IT specialists. *IEEE International Conference on Smart Information Systems and Technologies (SIST),* 1-5. [http://dx.doi.org/10.1109/SIST50301.2021.9465935]

Cobo, M.J., Martínez, M.A., Gutiérrez-Salcedo, M., Fujita, H., Herrera-Viedma, E. (2015). 25years at Knowledge-Based Systems: A bibliometric analysis. *Knowl. Base. Syst., 80*, 3-13. [http://dx.doi.org/10.1016/j.knosys.2014.12.035]

Durniak, A. (2000). Welcome to IEEE Xplore. *IEEE Power Eng. Rev., 20*(11), 12. [http://dx.doi.org/10.1109/39.883281]

Galvão, T.F. (2015). Principais itens para relatar Revisões sistemáticas e Meta-análises: A recomendação PRISMA. *Epidemiol. Serv. Saude, 24*(2), 335-342. [http://dx.doi.org/10.5123/S1679-49742015000200017]

Haddow, G. (2018). Bibliometric research methods. In Information, Systems, and Contexts (2nd ed., pp. 241–266). Chandos Publishing.

Hilmi, M. F., & Mustapha, Y. (2020). E-learning research in the Middle East: A bibliometric analysis. 2020 Sixth International Conference on e-Learning (econf), 243–248. [http://dx.doi.org/10.1109/econf51404.2020.9385513]

Hirsch, J.E. (2005). An index to quantify an individual's scientific research output. *Proc. Natl. Acad. Sci. USA, 102*(46), 16569-16572. [http://dx.doi.org/10.1073/pnas.0507655102] [PMID: 16275915]

Kraus, S., Breier, M., Dasí-Rodríguez, S. (2020). The art of crafting a systematic literature review in entrepreneurship research. *Int. Entrep. Manage. J., 16*(3), 1023-1042. [http://dx.doi.org/10.1007/s11365-020-00635-4]

Munikwa, S., & Mapara, J. (2022). Grappling with Education 5.0 curriculum development in Zimbabwean higher education: The case of Chinhoyi University of Technology. Journal of African Education, 199–200.

Nair, C.S., Patil, A., Mertova, P. (2009). Re-engineering graduate skills – a case study. *Eur. J. Eng. Educ., 34*(2), 131-139. [http://dx.doi.org/10.1080/03043790902829281]

S, O.M., J P, A. (2020). MapReduce and optimized deep network for rainfall prediction in agriculture. *Comput. J., 63*(6), 900-912. [http://dx.doi.org/10.1093/comjnl/bxz164]

PwC ASSOCHAM. (2024, July). Navigating the shift to Education 5.0: Enhancing higher education in India (pp. 4–7). Available from: https://www.assocham.org/uploads/files/Final%20Knowledge%20Report.pdf.

Sinha, S., Lacquet, B., & Maharaj, B. T. J. (2021). IEEE Xplore digital library indexes the Transactions of the SAIEE, 1909 to date. SAIEE Africa Research Journal, 112(4), 158–159.

Sina, L. B., & Nazemi, K. (2022). Visual analytics for systematic reviews according to PRISMA. 2022 26th

International Conference Information Visualisation (IV), 307–313.
[http://dx.doi.org/10.1109/IV56949.2022.00059]

Tang, S., Zhu, Q., Zhou, X., Liu, S., Wu, M. (2002). A conception of digital agriculture. *International Geoscience and Remote Sensing Symposium (IGARSS),* 3026-3028.
[http://dx.doi.org/10.1109/IGARSS.2002.1026858]

Tavares, M. C., Azevedo, G., Vale, J., & Bandeira, A. M. (2023). The accountant in the new era: Renewing the profession for Society 5.0. 2023 18th Iberian Conference on Information Systems and Technologies (CISTI), 1–6.
[http://dx.doi.org/10.23919/CISTI58278.2023.10211382]

Van Eck, N. J., & Waltman, L. (2019). VOSviewer manual (Version 1.6.13). Universiteit Leiden. Available from: https://www.vosviewer.com/documentation/Manual_VOSviewer_1.6.13.pdf.

Vieira, R., Monteiro, P., Azevedo, G., & Oliveira, J. (2023). Society 5.0 and Education 5.0: A critical reflection. 2023 18th Iberian Conference on Information Systems and Technologies (CISTI), 1–6.
[http://dx.doi.org/10.23919/CISTI58278.2023.10211386]

Yadav, S.A., Sahoo, B.M., Sharma, S., Das, L. (2020). An analysis of data mining techniques to analyze the effect of weather on agriculture. *2020 International Conference on Intelligent Engineering and Management (ICIEM),* 29-32.
[http://dx.doi.org/10.1109/ICIEM48762.2020.9160110]

Yong, L., Xiushan, L., Degui, Z., Fu, L. (2002). The main content, technical support and enforcement strategy of digital agriculture. *Geo Spat. Inf. Sci., 5*(1), 68-73.
[http://dx.doi.org/10.1007/BF02863497]

CHAPTER 5

New Approaches for Upskilling and Reskilling Digital Immigrants

Renji George Amballoor[1,*] and **Shankar B. Naik**[1]

[1] *Directorate of Higher Education, Alto-Porvorim, Goa, India*

Abstract: Higher Education Institutions (HEIs) are waging a war to address the skill gaps among their graduates, mainly digital immigrants. Many entry-level vacancies are not filled due to the absence of critical skills in graduates. At the same time, graduates complain about the lack of opportunities. The World Economic Forum realizes the gravity of this twin employment crisis and recommends innovative ideas for addressing the upskilling and reskilling emergency.

AI refers to computer systems that can perform tasks usually requiring human intelligence, such as learning, problem-solving, and decision-making. AI transforms industries by automating tasks, improving efficiency, and enabling new capabilities. AI may impact future jobs by creating new job categories, automating routine tasks, and enhancing decision-making processes. The increased use of disruptive automation, digital technologies, and artificial intelligence has necessitated new skills among graduates. The significant skill shift has been sped up by the Industrial Revolution, the spreading of the gig economy, and more. This shift means some jobs will disappear while new ones will emerge, requiring different skills. Therefore, there is a need for upskilling in AI and automation. Approaches for this include offering targeted training programs, integrating AI-related subjects into education, adopting multidisciplinary and interdisciplinary approaches, and encouraging continuous learning to help workers adapt to the changing job market.

This chapter aims to understand AI's significant impact on employment and the necessity of preparing the workforce for these changes. It also addresses the need and presents approaches for upskilling graduates who are digital immigrants.

Keywords: Digital immigrants, Higher education institutions, Reskilling, Skilling, Upskilling.

* **Corresponding author Renji George Amballoor:** Directorate of Higher Education, Alto-Porvorim, Goa, India; E-mail: amballoor@gmail.com

Rishikaysh Kaakandikar, Keshav Kaushik, Supriya Pathakm Madhavi Shamkuwar & Vivek Mishra (Eds.)

INTRODUCTION

The percolation of technology, the dreadful pandemic, and the increasing participation in green transition have created a new future of work scenarios. The nature of work has undergone tremendous changes with the increased use of digital technologies (World Bank, 2019). These technologies have created a new breed of platform-based business entrepreneurs, disrupting the existing input-output relations. The future of work is being increasingly discussed in local, national, and international forums. Two challenges are encountered in world economies: new entrants in the labor market do not possess the skills required by the industry, and the existing workforce is not equipped with the skills needed to survive in the future. Realizing the gravity of the situation, the World Economic Forum reports that the world economies are facing a reskilling-reskilling emergency.

World economies are experiencing substantial skill gaps among graduates, with their skills not compatible with labor market specifications. The right knowledge, skills, and abilities are significantly missing in our long queue of labor market graduates. Skill mismatch refers to vertical (levels of skills required) and horizontal (types of skills required) mismatches in terms of skill, skill gaps, skill shortages, and skill obsolescence (ILO). The deficits in the currently available skills to match the market requirements are mentioned as skill gaps.

McKinsey Global Institute surveyed young people and employers in 9 countries, including India, Brazil, Germany, the UK, *etc*. In the survey, 40% of employers identified the lack of skills as the main reason for entry-level job vacancies going unfilled. There is a twin employment crisis, *i.e*., a high level of youth unemployment simultaneously existing with a shortage of job seekers with critical skills required for the job opening (Sitek, 2012).

The mismatch between the skills of the currently available graduates and the skills required by the industry, along with the quality, is creating a significant challenge for policymakers (Mehrota *et al*., 2013). The skilling ecosystem includes higher education institutions, policymakers, enablers, implementing agencies, and beneficiaries (KPMG, 2014). According to KPMG (2016), the poor quality of higher education prevents graduates from obtaining suitable employment.

Depending on traditional ways to skill the graduates massively is nearly impossible, especially for a populous country like India. The demographic dividend is highly perishable and must be tapped effectively. The skilling ecosystem encounters many challenges. Benchmarking skilling programs to meet international standards and relevance quotient is a concern. In a VUCA (Volatility, Uncertainty, Complexity, and Ambiguity) world, ensuring program outcomes relevant to the changing world is impossible. Due to deficits in academia-industry linkages, Higher Education Institutions (HEIs) offering general programs do not have a first-hand understanding of the skill shifts and the emerging skill gaps. As a result, work-integrated learning opportunities like internships, industry projects, placements, *etc*., are denied to graduates. It will also result in the absence of experiential learning, the central pillar of skilling programs. In the traditional system, accessibility was a significant concern. Graduates have to travel long distances to enroll in training programs. The lack of transport and low frequency increased the problem of skilling graduates, especially those from remote villages—the lack of physical and digital infrastructure handicaps institutions in skilling graduates in large numbers. The poverty of master trainers in conducting skilling programs at multiple locations also added to the challenge. Due to a lack of technology, the trainers and the trainees cannot grasp the highly demanded skills.

The trillion-dollar question is optimizing the skilling program for the broadest reach, impact, and effectiveness. Brick and mortar strategy of skill training will require a very long period, a vast number of resources, and lots of workforce to ensure the tapping of dividends. Higher Education Institutions (HEIs) have a significant role in not only skilling the graduates but also offering upskilling and reskilling opportunities for the workforce. Artificial Intelligence (AI) and its bandwagon of automation tools are a solution for the skilling, upskilling, and reskilling needs. Graduates undertaking learning to expand their existing skill sets is known as upskilling. Learning a new set of skills not mandated by the syllabi and the market requirements is reskilling.

THE FUTURE OF JOBS IN THE AGE OF AI

One of the most significant changes we will see as technology advances is in the job market. This is mainly due to the rise of Artificial Intelligence (AI). AI has already become a part of our everyday lives. AI shows its presence, from the virtual assistants on our phones to the algorithms that suggest what we should watch next on streaming platforms. However, its influence on jobs in the future will be even more significant (Poba-Nzaou, 2021).

One of the most exciting things about AI is that it will create entirely new types of jobs that we can't even fully imagine now. As the internet gave rise to jobs like social media manager or app developer, AI will lead to new roles like AI trainers, who teach machines how to improve their performance, or AI ethics officers, who ensure that AI systems are used responsibly (Hunt, 2022). These jobs will require a mix of technical knowledge and creative thinking. This means people must be open to learning new skills throughout their careers.

In the future, technical skills like coding or data analysis will not suffice. Many jobs will require a combination of skills. For example, someone might need to know how to analyze data and also be able to communicate their findings clearly to others. Creativity and problem-solving will also be highly valued, especially in roles where AI is used to support human work rather than replace it.

AI technology is constantly changing and improving. Because of this, workers will need to keep learning throughout their lives. Learning a skill once and using it for the entire career might not work. Instead, people must regularly update their skills to keep up with new tools and technologies. This means that schools and universities must also adapt, focusing on teaching students how to learn rather than just teaching specific skills.

It is important to remember that AI will also work alongside us. This will make many jobs easier and more efficient. For instance, in industries like healthcare, AI can help doctors by quickly analyzing medical images or predicting patient outcomes. This doesn't replace the doctor but gives them more tools. In many fields, AI will help people do their jobs better rather than taking those jobs away.

Jobs that require emotional intelligence, complex decision-making, or human interaction will still be important. For example, jobs in healthcare, education, or creative industries like writing and the arts will be harder for AI to replace. AI is excellent at processing information and finding patterns, but it is not as good at understanding emotions, thinking creatively, or making complex ethical decisions. This means that jobs that require these skills will be in demand.

One way AI might change the job market is by making work more flexible. AI tools can allow people to work from anywhere, as long as they have an internet connection. This could lead to more remote work opportunities, where people don't need to be in an office to do their jobs. It also means that companies might hire people from all over the world, not just from their local area. This could be a great opportunity for people in regions where jobs are harder to find.

As AI continues to evolve, jobs are likely to change more frequently (Acemoglu, 2022). This means that people might have shorter careers in one specific role before moving on to something new. While this can be challenging, it also means that there will be opportunities to try out different types of work and learn new things. Workers will need to be adaptable and open to change, which could lead to more varied and interesting career paths.

The gig economy, where people work on short-term projects or freelance, is already growing. AI is likely to accelerate this trend (George, 2023). AI platforms can match workers with jobs more quickly and efficiently, which could create more opportunities for freelance work. This might be particularly appealing for people who prefer the flexibility of when and where they work. However, it could also mean that more people will have to deal with the challenges of gig work, such as job insecurity and lack of benefits.

AI has the potential to make workplaces much more efficient by automating routine tasks and streamlining operations. For example, AI could handle tasks like scheduling, data entry, or even customer service inquiries, allowing workers to focus on more complex and creative tasks. However, this increase in efficiency could also lead to fewer jobs in some areas, especially in roles that involve repetitive tasks.

As AI becomes more integrated into our lives, there will be a growing need for jobs related to overseeing and managing AI systems (Muda, 2023). These roles will involve ensuring that AI is used ethically, complies with laws and regulations, and doesn't harm people or society. This could create new opportunities in fields like law, policy, and ethics, as well as in technical roles focusing on AI safety and reliability.

One concern about the rise of AI is that it could lead to greater income inequality. High-paying jobs that require specialized AI skills become more common, while lower-paying jobs that are easier to automate could disappear. This could create a wider gap between those with the skills to work with AI and those without. Addressing this issue will be essential to ensure that the benefits of AI are shared more equally (Oluwaseyi, 2024).

In the future, many jobs will require knowledge across multiple fields. For example, someone working in finance might need to understand both the financial industry and how AI can be used to predict market trends. Similarly, a healthcare worker might need to know medicine and how AI tools can help diagnose diseases. This means that people who combine knowledge from different areas will be highly valued in the job market.

As AI changes the workplace, leadership roles will also evolve. Leaders must understand how to manage teams that include people and AI systems. They must also focus on strategic thinking, collaboration, and ethical decision-making. This could change the way companies are run, with a greater emphasis on adaptability and innovation.

While AI will create new opportunities, it will also displace some jobs, particularly in industries that rely on manual or repetitive tasks. Many workers will need to learn new skills or transition to different roles. Governments, companies, and educational institutions will need to help people reskill and find new employment opportunities.

As AI takes over more routine tasks, jobs that rely on creativity, originality, and innovation will become more critical. Whether in the arts, product design, or marketing, thinking creatively will be a key asset in the job market. AI can assist with data and analysis, but human creativity is something that is harder to replicate.

The rise of AI will require changes in education. Schools and universities must focus more on teaching students how to learn and adapt rather than just preparing them for specific jobs. This might include more emphasis on critical thinking, problem-solving, and collaboration. Lifelong learning will also become more important as people must continue developing new skills throughout their careers (Udvaros, 2023).

Examples of Successful Government and NGO Interventions for Reskilling Digital Immigrants –

Singapore's SkillsFuture Initiative

Type: Government
Focus: Lifelong learning for all citizens, including older workers and digital immigrants.

Details:

- Provides training credits to all adults, regardless of employment status.
- Includes AI, data analytics, and digital marketing modules.
- Offers career advisory services and employer-sponsored courses.
  Impact: Built a nationwide culture of lifelong learning and helped thousands of older professionals re-enter tech-savvy roles.

IBM SkillsBuild

Type: NGO–Corporate Collaboration
Focus: Tailored digital training for underserved communities

Details:

- Provides free access to online courses on digital and workplace skills.
- Collaborates with local schools, municipalities, and NGOs.
- Specifically targets youth, veterans, and adults unfamiliar with tech (*i.e.*, digital immigrants).

Impact: Supported thousands globally in transitioning into digital roles through mentoring and structured curricula.

Germany's Dual Education System

Type: Government
Focus: Blends academic education with practical on-the-job training

Details:

- Used as a reskilling model for both youth and mid-career workers.
- Encourages apprenticeships in digital industries like robotics and manufacturing.
- Supported by German federal and regional governments.
  Impact: One of Europe's most successful workforce reintegration models for mid-career upskilling.

European Union's Digital Skills and Jobs Coalition

Type: Supranational/Government Partnership
Focus: Reskilling the European workforce, especially older populations

Details:

- Brings together industry, NGOs, and educators.
- Provides regional hubs, digital literacy training, and awareness campaigns.
- Prioritizes inclusion of women, elderly workers, and rural populations.
  Impact: A leading platform addressing the digital divide in Europe

THE NEED TO UPSKILL IN AI

Technology is advancing rapidly in today's fast-changing world, and Artificial Intelligence (AI) is at the forefront of these changes. AI is everywhere, from smartphones to workplaces, and it will only become more critical. Because of this, it's crucial for everyone, whether a student, a professional, or someone simply interested in staying current, to upskill in AI (Autor, 2024).

One of the main reasons we need to upskill in AI is that the technology is constantly evolving. What was cutting-edge a few years ago might now be outdated. New tools, techniques, and applications of AI are being developed all the time. If we don't keep learning, we risk falling behind. This can make staying competitive in the job market harder because employers are looking for up-to-date people with the latest technologies. We can ensure we're always ready to take on new challenges and opportunities by continuously learning and upskilling in AI.

AI is likely to replace jobs, especially those that involve repetitive tasks. This is a genuine concern for many people. However, by upskilling in AI, we can secure jobs and even open up new career opportunities. Instead of being replaced by AI, we can learn how to work with it (Zang, 2024). For example, someone who understands AI might move from a routine job to a more advanced role where they use AI to make decisions or solve problems. By learning new AI skills, we make ourselves more valuable to employers, which can lead to greater job security.

AI is not just about keeping the job we have. It is also about advancing in our careers. Many of the highest-paying and most exciting jobs today are in AI-related fields, such as data science, machine learning, and robotics. By upskilling in AI, we can take advantage of these new opportunities. Learning AI can help us get promotions, switch to more interesting roles, or even start our businesses. The demand for AI skills is growing, and those with these skills can often command higher salaries and enjoy greater job satisfaction.

AI has the power to make our work easier and more efficient. By learning to use AI tools, we can automate tasks that take time and effort. For instance, AI can help us analyze large amounts of data quickly, allowing us to make better decisions faster. AI is used in healthcare, finance, marketing, and more. By upskilling in AI, we can learn how to integrate these tools into our daily work, making us more productive and effective in our roles.

As AI becomes more integrated into various industries, job competition is also changing. Employers are increasingly looking for candidates who have AI skills, even in fields where AI wasn't traditionally used. For example, knowing how to use AI to analyze customer data in marketing can give someone a significant edge over others. By upskilling in AI, we can stay competitive in the job market, ensuring we have the skills employers seek.

AI is transforming industries like manufacturing, finance, healthcare, and retail. The skills required to succeed in these fields also evolve as these changes happen. By upskilling in AI, we can adapt to these changes and thrive in our careers. For instance, in manufacturing, AI is used for everything from managing supply chains to predicting equipment failures. Workers who understand AI are better positioned to assume new roles and responsibilities as their industry evolves.

Rather than seeing AI as a threat, it's better to view it as a collaborator. AI can handle much of the heavy lifting, like processing data or performing routine tasks, allowing us to focus on more creative and strategic work. However, we must understand how it works to collaborate effectively with AI. Upskilling in AI helps us learn how to make the most of these powerful tools, leading to better results and more innovative solutions in our work.

As AI becomes more powerful, it's essential to use it responsibly. This includes understanding AI's ethical implications, such as privacy, bias, and fairness issues. By upskilling in AI, we can contribute to developing and using AI to benefit society. This is especially important for those in leadership or decision-making roles, who must make informed choices about how AI is implemented.

Upskilling in AI is essential for staying relevant, securing our jobs, advancing our careers, and adapting to the rapid changes brought about by technology. It's not just about learning new skills. It is about preparing ourselves for the future and ensuring we can work effectively with AI to create better outcomes in whatever field we are in (Brawner, 2023).

APPROACHES TO UPSKILLING IN AI

As AI grows in importance across various industries, learning how to use these technologies effectively is essential. Whether starting a career, looking to advance in a current job, or considering a career change, there are many ways to upskill in AI.

One of the easiest and most accessible ways to start learning AI is through online courses. Many platforms offer courses specifically focused on AI. These courses range from beginner levels, where one can learn the basics of AI, to more advanced levels that dive deep into specific areas like machine learning, data science, and natural language processing. Online courses are convenient because one can learn independently, fitting the studies around the schedule. Many of these platforms also offer certifications upon completion of the course. Earning a certification can be a great way to show potential employers that one has the skills they are looking for.

Enrolling in a university program might be the best option for those who want a more in-depth understanding of AI. Many universities now offer degrees in AI, Data Science, and related fields. These programs provide a solid foundation in AI concepts and hands-on experience with the tools and techniques used in the industry. University programs often include access to resources like labs, research opportunities, and internships, which can be valuable for gaining practical experience. Additionally, having a degree in AI or a related field can open up more job opportunities. It can be particularly beneficial if one is looking to enter a specialized area of AI. Many institutions now offer online degrees for those who cannot attend a traditional university. These programs provide the same education as their on-campus counterparts but with the flexibility to study from anywhere globally.

For a more intense and focused learning experience, AI boot camps could be a great option. Bootcamps are short, immersive programs that teach practical AI skills in weeks or months. They are designed to get participants job-ready quickly, with a strong emphasis on hands-on learning. During an AI boot camp, participants typically work on real-world projects that allow them to apply their knowledge. This is a great way to build a portfolio to show potential employers. Bootcamps often connect with companies looking to hire AI talent, which can be a significant advantage when one is ready to start the job search.

For those who prefer learning in shorter bursts or want to focus on specific AI topics, workshops and seminars can be a valuable resource. These are often one-day or weekend events where experts in the field teach a particular aspect of AI. For example, one might attend a workshop on how to use a specific AI tool or a seminar on the ethical considerations of AI. Workshops and seminars are a great way to gain new skills quickly without a long-term commitment. They also provide an opportunity to meet other people interested in AI, which can be helpful for networking and learning from others.

One of the most exciting ways to upskill in AI is by participating in AI competitions and hackathons. Platforms like Kaggle host AI competitions where participants work on solving real-world problems using AI. These competitions are open to everyone, from beginners to experts, offering a practical way to apply what one has learned. Hackathons are similar but often occur over a shorter period, like a weekend. During a hackathon, teams work together to build AI projects from scratch. This is a fun and challenging way to develop AI skills, and it can also lead to opportunities for prizes, recognition, and even job offers.

For those who prefer learning independently, reading books, articles, and research papers on AI can be a powerful way to deepen their knowledge. Many excellent books on AI cover everything from the basics to more advanced topics. Research papers can also help to stay up-to-date with the latest developments in AI. In addition to books and papers, many online resources are available, including blogs, podcasts, and video tutorials. These resources can provide insights into current trends in AI and offer practical tips for applying AI in various fields. The key to self-study is to be disciplined and consistent, setting aside regular time to learn and practice.

One of the best ways to learn AI for those already working is through on-the-job training. Many companies are incorporating AI into their operations, and there may be opportunities to learn new AI skills as part of their current role. For example, if one works in marketing, one might learn how to use AI to analyze customer data and improve campaigns. On-the-job training can be efficient because participants learn in a real-world context, applying AI to the company's actual problems. If the job doesn't currently involve AI, one can consider talking to the employer about how to get involved in AI projects. Many companies are willing to invest in upskilling their employees, especially in such a critical area.

Another great way to upskill in AI is by working on personal projects. This involves choosing a project, such as creating a chatbot, building a recommendation system, or analyzing data. Then, AI tools can be used to bring it to life. Personal projects allow us to apply our skills in a meaningful way and can help us build a portfolio to showcase our abilities. Working on individual projects will also enable me to experiment and learn from mistakes without the pressure of deadlines or external expectations. One can take time, explore different approaches, and better understand how AI works.

Learning doesn't have to be a solitary activity. One can connect with others and learn AI by joining AI communities and forums. These communities are often full of people eager to share their knowledge and help each other. For example, one might join a Reddit group focused on machine learning or a LinkedIn group for AI professionals. In these communities, one can ask questions, share experiences, and collaborate on projects. Being part of a community can provide support, motivation, and inspiration as one continues to learn.

AI podcasts and webinars can be an excellent resource for those who like to learn on the go. Many podcasts cover AI topics, from basic introductions to in-depth discussions with industry experts. Listening to podcasts can be a convenient way to learn during commuting, workouts, or downtime. Webinars are typically live events where one can learn about specific AI topics and ask questions. Many organizations offer free webinars that cover the latest trends and developments in AI, making them an excellent way to stay informed and continue learning.

Continuous learning is essential to keep up with the fast-paced world of AI. Some platforms offer subscriptions that provide access to a wide range of AI learning resources. These platforms often provide structured learning paths that guide users through different levels of AI knowledge, from beginner to advanced. One can always be up-to-date with the latest AI tools and techniques by subscribing to a continuous learning platform. This approach benefits those who want to keep their skills sharp and ready for whatever comes next in AI. By exploring these various AI upskilling approaches, one can choose the ones that best fit their learning style, goals, and schedule. Whether taking online courses, reading books, or working on personal projects, the important thing is to keep learning and growing AI skills.

With the right approach, one will be well-prepared to take advantage of AI's many opportunities.

ROLE OF HEIS IN UPSKILLING GRADUATES IN AI

Higher Educational Institutions (HEIs) can play a crucial role in helping graduates, especially digital immigrants, acquire the necessary AI skills to succeed in future jobs (Padmaja, 2021). One practical approach is to create specialized AI courses and certification programs. These courses should start with basic concepts and gradually move to more advanced topics. By doing so, digital immigrants, who may not be as familiar with technology, can build a strong foundation before tackling more complex material (Murdan and Halkhoree, 2024) (Aithal and Maiya, 2023). Completing these programs can also provide students with recognized credentials that enhance their job prospects.

Another way to support learning is through AI-focused workshops and boot camps. These short-term, intensive programs offer hands-on learning experiences that allow students to deeply engage with practical AI applications (Awasthy *et al.*, 2023). For digital immigrants, this immersive method makes learning more engaging and helps build confidence as they apply AI concepts in real-world situations.

Mentorship programs can also be highly beneficial. In these programs, students are paired with experienced professionals or faculty members with AI expertise. Mentors offer personalized guidance, answer questions, and encourage, making learning less overwhelming for those new to the technology.

AI literacy programs are another essential approach, especially for non-technical business, humanities, or social sciences students. These programs focus on helping students understand the basics of AI, its ethical implications, and how it can be applied in their specific fields. This ensures that digital immigrants in these areas are not left behind and can see how AI might impact their future careers.

Flexibility in learning options is also key to accommodating the diverse needs of digital immigrants. HEIs can offer online courses, evening classes, or part-time programs that allow students to learn at their own pace. This flexibility is essential for students with other commitments, such as work or family responsibilities, ensuring that AI education is accessible to everyone.

Collaborating with industry is a powerful way for HEIs to keep their AI programs up-to-date and relevant. Partnerships with companies and industry leaders can lead to guest lectures, internships, and access to the latest AI tools and technologies. By working closely with industry, HEIs can provide students with practical skills directly applicable to the job market, helping them transition smoothly into AI-related roles.

Data-driven decisions in Higher Education Institutions (HEIs) can help personalize learning according to students' needs and capabilities. Tailored recommendations can assist educators in providing unique learning experiences to every student, ultimately attracting higher enrollment and positioning ahead of competitors.

In a data-driven decision process, educators can identify hidden patterns in the data and support forecasting and predictive analysis for better decision-making.

Data-driven decisions make necessary data accessible to every segment of the institution, eliminating the data monopoly held by a few individuals or departments. The democratization of data will encourage creative and innovative interventions in HEIs (Ehlinger and Stephany, 2023).

CONCLUSION

The rapid advancement of technology, the impact of the pandemic, and the shift towards green initiatives have dramatically changed how we work. Many graduates and workers are unprepared for modern job requirements due to significant skill gaps. Traditional education methods are insufficient to address these challenges, especially in countries with large populations like India. The World Economic Forum has highlighted this as a "reskilling emergency."

Higher Educational Institutions (HEIs) are vital in solving this problem. By embracing innovative approaches, such as integrating Artificial Intelligence (AI) into their programs, they can help bridge the skill gap. AI offers flexible and scalable solutions for skilling, upskilling, and reskilling many graduates. **Approaches like offering specialized AI courses, organizing practical workshops,** setting up mentorship programs, providing AI literacy for non-technical fields, offering flexible learning options, and collaborating with industries can make a significant difference.

These strategies can help graduates, including those less familiar with technology, often called digital immigrants, to gain the necessary skills for today's job market. By focusing on practical experience, making learning accessible, and ensuring that education is relevant to industry needs, HEIs can prepare students to meet the evolving demands of the workforce. This not only benefits individual careers but also contributes to the economic growth and competitiveness of the country. Embracing these methods is essential to make the most of the available talent and to overcome the challenges of a rapidly changing work environment.

REFERENCES

Acemoglu, D., Autor, D., Hazell, J., Restrepo, P. (2022). Artificial intelligence and jobs: Evidence from online vacancies. *J. Labor Econ., 40*(S1), S293-S340.
[http://dx.doi.org/10.1086/718327]

Aithal, P.S., Maiya, A.K. (2023). Innovations in higher education industry–Shaping the future. *International Journal of Case Studies in Business, IT, and Education, 7*(4), 294-322. [IJCSBE].
[http://dx.doi.org/10.47992/IJCSBE.2581.6942.0321]

Awasthy, R., Tegabala, S., John, B., Alsamarra'i, Z., Kovalev, A., Dilnutt, R. (2023). Enhancing graduate skills through bootcamps. *Journal of Higher Education Development, 10*(3), 213-222.
[http://dx.doi.org/10.54355/jhed.2023.10.3.213]

Autor, D. (2024). Applying AI to rebuild middle-class jobs (No. w32140). *National Bureau of Economic Research.*
[http://dx.doi.org/10.3386/w32140]

Brawner, K., Wang, N., Nye, B. (2023). Teaching artificial intelligence (AI) with AI for AI applications. *The International FLAIRS Conference Proceedings, 36*, 64-72.
[http://dx.doi.org/10.32473/flairs.36.133388]

Ehlinger, E. G., & Stephany, F. (2023). Skills or a degree?: The rise of skill-based hiring for AI and green jobs. Bruegel. Availale from: https://www.bruegel.org/skills-or-degree.

George, A.S., George, A.H., Martin, A.G. (2023). ChatGPT and the future of work: A comprehensive analysis of AI's impact on jobs and employment. *Partners Universal International Innovation Journal, 1*(3), 154-186.

Hunt, W., Sarkar, S., Warhurst, C. (2022). Measuring the impact of AI on jobs at the organization level: Lessons from a survey of UK business leaders. *Res. Policy, 51*(2), 104425.
[http://dx.doi.org/10.1016/j.respol.2021.104425]

KPMG. (2014). Skilling India: A look back at the progress, challenges, and the way forward. KPMG.

KPMG. (2016). Re-engineering the skill ecosystem. KPMG.

Mehrota, S., Gandhi, A., & Sahoo, B. K. (2013). Estimating the skill gap on a realistic basis for 2022. Institute of Manpower Research, Planning Commission.

Muda, I., Almahairah, M.S., Jaiswal, R., Kanike, U.K., Arshad, M.W., Bhattacharya, S. (2023). Role of AI in decision making and its socio-psycho impact on jobs, project management and business of employees. *Journal for ReAttach Therapy and Developmental Diversities, 6*(5s), 517-523.
[http://dx.doi.org/10.26407/2023.6.s5.559]

Murdan, A.P., Halkhoree, R. (2024). Integration of artificial intelligence for educational excellence and innovation in higher education institutions. *1st International Conference on Smart Energy Systems and Artificial Intelligence (SESAI),* 1-6.
[http://dx.doi.org/10.1109/SESAI61023.2024.10599402]

Oluwaseyi, J., Cena, J. (2024). Analyzing the impact of artificial intelligence on jobs. *Statistics, 14*(1), 150-155.
[http://dx.doi.org/10.16905/14.1.150]

Padmaja, V., Mukul, K. (2021). Upskilling and reskilling in the digital age: The way forward for higher educational institutions. *Transforming higher education through digitalization.* CRC Press.
[http://dx.doi.org/10.1201/9781003132097-15]

Poba-Nzaou, P., Galani, M., Uwizeyemungu, S., Ceric, A. (2021). The impacts of artificial intelligence (AI) on jobs: an industry perspective. *Strategic HR Rev., 20*(2), 60-65.
[http://dx.doi.org/10.1108/SHR-01-2021-0003]

Sitek, S. (2012). Bridging the skill gap: Help wanted, skills lacking: Why the mismatch in today's economy? American Society for Training & Development (ASTD).

Tariq, M.U. (2024). The role of AI in skilling, upskilling, and reskilling the workforce. *Integrating generative AI in education to achieve sustainable development goals.* IGI Global. [http://dx.doi.org/10.4018/979-8-3693-2440-0.ch023]

Udvaros, J., Forman, N. (2023). Artificial intelligence and Education 4.0. *INTED2023 Proceedings,* 6309-6317. [http://dx.doi.org/10.21125/inted.2023.1674]

World Bank. (2019). The changing nature of work. World Bank.

Zang, H., Li, S., Dong, X., Ma, D., Dang, B. (2024). Evaluating the social impact of AI in manufacturing: A methodological framework for ethical production. *Academic Journal of Sociology and Management,* *2*(1), 21-25.

CHAPTER 6

AI-Driven Automation for Upskilling and Reskilling the Workforce: An ISM and MICMAC Approach

Sushma Sharma[1], **Anju Rohilla**[1,*] and **Priya Jindal**[2]

[1] *Department of Management Studies, SRM University, Delhi-NCR, Sonepat, India*

[2] *Chitkara Business School, Chitkara University, Punjab, India.*

Abstract: The advancement of artificial intelligence and automation technologies is changing workforce dynamics, necessitating new methods for employee upskilling and reskilling. This study emphasises AI-driven automation for labour force development, applying Interpretive Structural Modeling and MICMAC analysis to develop a planned framework for developing skills in organizational settings. Key factors identified in the analysis include Skill Identification and Gap Analysis, Employee Training Programs, Technological Integration, Leadership Support, Resource Allocation, Employee Engagement, Learning and Development Infrastructure, Organizational Culture, Performance Measurement, and Policy and Regulatory Support. By analysing 56 articles published between January 1990 and December 2023, this research highlights gaps in the literature and suggests a planned roadmap for further research and real-world applications. The findings highlight AI's potential to achieve sustainable competitive advantage, improve employee performance by developing a continuous learning environment, essential for addressing the challenges posed by automation and technological advancements.

Keywords: AI Automation, Artificial intelligence, Employee engagement, Employee training programs, Gap analysis, ISM, Learning and development infrastructure, Management support, MICMAC, Policy and regulatory support, Resource allocation, Skill identification, Technological integration, Upskilling.

INTRODUCTION

The advancement of artificial intelligence and automation is changing the global workforce, primarily restructuring the skills desired to succeed in today's industries. AI is defined as a computer program that can learn and emulate human

* **Corresponding author Anju Rohilla:** Department of Management Studies, SRM University, Delhi-NCR, Sonepat, India; E-mail: anjurohilla@srmuniversity.ac.in

behaviour with the ultimate goal of becoming more intelligent than humans (Kaplan and Haenlein, 2019). With the inception of industry 4.0 and 5.0, businesses are integrating AI in their operation from the production of goods and services to customer relation management. Even the integration of AI in businesses leads to the development of innovative and unique business models, with Fintech and Insurtech being prime examples. But at every stage of the change and development, human resources are the crucial aspect that every business should consider. A more innovative and adaptable workforce is essential for successfully implementing an innovative business model or bringing about any strategic change. On the other hand, the organizations must employ their human resourced more precisely and effectively. It directly results in the use of robotic process automation, and HR analytics become vital for the organizations.

Organizations are finding opportunities as well as challenges due to the incorporation of AI in HRM. In the era of Industry 4.0, where AI has become more prevalent and technological advancements encourage top executives to reconsider their strategies for developing their human resources to meet today's world requirements (Dwivedi *et al.*, 2023). Artificial intelligence has a significant impact on the labour market as it has transformed the industry to an AI-driven, automated industry from the tradition mechanization industry (Acemoglu, 1998). It is anticipated by specialists that machine learning and artificial intelligence would bring about a substantial transformation in the character of work in the next decade (Rahman & Abedin, 2021; Tommasi *et al.*, 2021). For instance, PathAI, which aids in cancer diagnosis, and chatbots used in customer service; these use Natural Language Processing (NLP) to mimic human conversations and instantly answer customer questions.

In the late 1990s and early 2000s, experts recognized the potential of AI in increasing labour productivity through automation (Brynjolfsson and Hitt, 1998; Levy and Murnane, 2004). On the contrary, late 2020s research shows that AI will substitute humans in decision-making (Russell and Norvig, 1995). AI can automate many human-performed jobs; its widespread use will affect both white-collar and blue-collar employees (Leinen *et al.*, 2020). This perspective holds that while this shift might potentially boost efficiency and productivity for knowledge workers, it could also cause them to lose their jobs. One in fourteen workers worldwide, or 375 million people, could have to find new employment opportunities as a result of AI-related technology developments by 2030.

Many physical labour-intensive jobs may become obsolete as a result of AI, which would have a disproportionately negative effect on blue-collar workers. Manual, operational, and visual-spatial abilities, which are considered middle skills, are seeing a decline in demand. Contrarily, others claim that AI's incorporation into the workplace may actually result in the addition of new positions, particularly in industries that are heavily invested in AI research and development and implementation (Puzzo *et al.*, 2020). To remain competitive, organizations must focus on reskilling their employees, while the workforce should be proactive in upskilling themselves (Hancock *et al.*, 2020).

Upskilling is the process of acquiring new skills or imparting such talents to employees so that they can do their current jobs better. Conversely, reskilling comprises learning new abilities to move into a different work function or preparing people for changing occupations. In the next five years, there will be a 40% shift in the distribution of essential skills, and half of the workforce will need some kind of re-education or training (World Economic Forum [WEF], 2020).

AI has the potential to increase social inequality (Levy, 2018) and eliminate more than 45% of all occupations (Berg, Buffie, and Zanna, 2018; Leinen *et al.*, 2020). There are potential upsides as well, such as upgrading the jobs, creating new job opportunities rather than displacing occupations (*e.g.*, Autor, 2015). These two-faced sides of AI underlie the need for balanced measures that address the possibilities of job losses while simultaneously offering possibilities *via* targeted upskilling and reskilling efforts. Artificial Intelligence has become more and more important in HRM as more and more companies adopt remote and hybrid work models due to the epidemic of Covid-19 (Nguyen and Malik, 2021; Malik *et al.*, 2022). The use of AI-driven automation in HRM operations such as talent development, performance management, and recruitment and selection has been acknowledged by many researchers (Saukkonen *et al.*, 2019; Vrontis *et al.*, 2021), but how AI may derive the initiatives to upskill and reskill the employees is still unexplored. However, a gap is seen in the literature about how AI may help with strategic workforce development, specifically in relation to multilevel HR frameworks and upskilling and reskilling (Jaiswal *et al.*, 2022).

This chapter seeks to fill the gap that how AI-driven automation might encourage the reskilling and upskilling of the workforce. It also provides insights into how organizations may harness Artificial Intelligence to prepare their workforce for the opportunities and challenges posed by AI-driven technological advancements by exploring the progress of AI from 1995 to 2023 and its impact on skill development.

This chapter is divided into eight sections. The first section introduces Artificial Intelligence and the need for upskilling and reskilling; the second section defines the objective of the study conducted in this chapter, followed by explaining the research methodology—Interpretive Structural Modeling (ISM)—in the third section. The Fourth section identifies the variables rooted in the literature, and the fifth section addresses the gap. Results of the research study are discussed in the sixth section, followed by recommendations and suggestions in the next section. The last section of the study provides the limitations and policy implications.

OBJECTIVE OF THE STUDY

The conducted study is aimed at investigating the strategic application of AI-driven automation to encourage the upskilling and reskilling of the workforce. This research intends to build a multilevel theoretical framework through applying the Interpretive Structural Modelling and MICMAC analysis by identifying the key variables, such as organizational culture, technical integration, managerial support, skill gap analysis, and staff training program. This model will examine the interdependencies and driving forces between these key variables. Additionally, the study will seek to identify the research gaps by investigating AI's impact on workforce development and offer implications and suggestions to the industry that how they can integrate AI successfully in their organizations. Furthermore, this study will provide a strategic roadmap to improve employee performance and productivity, encourage learning, and boost creativity in employees.

RESEARCH METHODOLOGY

The research methodology applied in this study is based on Interpretive Structural Modeling (ISM), a tool introduced by Warfield in 1976, which is considered to evaluate the relationships and interactions among diverse variables. ISM is an active approach for understanding complexities by identifying and categorising key variables that influence a particular research domain. In this study, ISM is employed to analyse the factors linked with AI-driven automation for upskilling and reskilling the workforce, focusing on their interrelationships and hierarchical structures.

IDENTIFICATION OF VARIABLES

Variables are identified through a review of literature and expert views to evaluate the recent trends in AI-driven automation for upskilling and reskilling the workforce. Numerous studies have provided insights into the use of Artificial Intelligence (AI) in developing human resources. However, prior research has focused either on the technical aspects of AI or its implications, without fully integrating both domains. This study aims to highlight this gap by analysing the combination of AI applications in upskilling and reskilling of employees, guaranteeing a complete approach to workforce development in the context of AI-driven automation.

After reviewing literature from January 1990 to December 2023 and accompanying discussions with experts in the field of AI and skill enhancement, the following critical factors were identified as key drivers for the current study.

Skill Identification and Gap Analysis

Identification of skills is central to aligning workforce competencies with organizational objectives. (Jackson and Schule, 2003) stated that comprehensive skills assessments allow organizations to classify current skill gaps, which AI can automate for greater proficiency. AI-driven platforms can analyse employee performance and recommend specific upskilling paths based on gaps. (Lee and Lee, 2020) found that AI-powered learning platforms, such as LMS (Learning Management Systems), help HR experts tailor training programs based on skill gap assessments, leading to more effective learning outcomes.

Employee Training Programs

AI enables the development and delivery of training programs that cater to individual employee needs. (Hager and Gonczi, 1996) recommended that tailored learning is more effective than traditional training approaches, and AI can systematize this personalization. Chauhan *et al.* (2020) observed the role of AI in designing adaptive learning environments, which adjust to employees' learning progress, guaranteeing continuous skill development and advanced engagement levels.

Technological Integration

Combining new technology in the workplace can enhance operational efficiency and employee capabilities equally. Kohli and Melville (2019) assessed how the use of technology in an organisation is directly linked with employee performance. AI-driven automation contributes a lot to speeding up this integration, making technological tools more friendly for employees. A study by Rao and Prasad (2022) showed that employees using AI-enhanced tools practiced skill acquisition, improving overall job performance.

Management Support and Leadership

Strong leadership is a key requirement for the application of AI-driven upskilling and reskilling initiatives. Numerous studies have proved the status of management support in employee development efforts. For instance, Stoker *et al.* (2019) emphasized that leadership plays a crucial role in motivating employees to adopt AI technologies. Additionally, Saleh *et al.* (2021) found that transformational leadership, in particular, is significant in inspiring employees to get involved in upskilling initiatives driven by AI technologies.

Resource Allocation

The use of adequate resources, both financial and technological, is crucial for AI-enabled workforce development initiatives. Sharma and Mishra (2019) considered how resource constraints can delay AI adoption for training drives in organizations. Sufficient resource allocation guarantees that employees receive access to the latest AI tools and training programs, thus empowering effective upskilling and reskilling initiatives. Singh *et al.* (2021) proved that organisations participating more in AI technologies observe higher productivity gains over their reskilled workforce.

Employee Engagement and Motivation

AI has been shown to certainly affect employee engagement by offering personalized learning and career development options. According to Alfes *et al.* (2013), highly engaged employees are more likely to contribute to skill development programs. Dissanayake *et al.* (2023) stated that the application of AI-driven performance feedback systems suggested increased employee enthusiasm to be involved in reskilling activities. AI's capacity to offer real-time feedback and career opportunities can strengthen a strong culture of learning and development.

Learning and Development Infrastructure

A healthy learning infrastructure system is significant for applying AI-driven upskilling programs. Studies have proven that organizations with strong learning environments are more effective in facilitating continuous learning. Garavan *et al.* (2016) noted the importance of learning infrastructure in guaranteeing employees stay current with technological changes. AI plays a significant role in powering this infrastructure, offering scalable, on-demand learning resources. A study by Bhardwaj and Kumar (2022) proved that AI-based learning systems pointedly reduced the time needed to acquire new skills compared to traditional training methods.

Organizational Culture

Organizational culture that encourages innovation and continuous learning can influence the success of AI-driven upskilling initiatives. A study conducted by Cameron and Quinn (2011) recommends that an ethos of learning and technology implementation is crucial for the success of AI-driven transformation. A recent study by Silva *et al.* (2021) presented that companies with a culture of AI implementation saw higher levels of employee participation in upskilling programs, nurturing innovation, and competitiveness.

Policy and Regulatory Support

Regulatory and policy frameworks surrounding AI and workforce development play a key role in shaping how organizations approach upskilling and reskilling. A study by Caldwell (2015) emphasized that supportive policies at the national and organizational levels promote more widespread AI adoption. With ethical concerns and job displacement worries rising, regulatory support ensures that AI-driven reskilling hard work is aligned with employee rights and labour laws. Carvalho and Abraham (2022) conferred how government policies aimed at promoting AI literacy and ethical deployment in workforce training have developedessential to the success of AI-driven HR strategies.

GAP OF THE STUDY

Despite the growing body of research on AI's automation and workforce development, there is a notable gap in studies that combine technical aspects of AI-driven upskilling and reskilling. Most existing research focuses either on the operational benefits of AI in HRM or on ethical concerns in isolation. However, little attention has been given to a holistic framework that addresses both the technological implementation and ethical considerations simultaneously. Additionally, the use of methodologies like ISM and MICMAC to analyse the interdependencies among variables influencing AI-driven workforce development remains underexplored. This study aims to fill these gaps by offering a comprehensive, multilevel approach to understanding AI's role in strategic workforce development.

Variables used for Model Construction

Table **1** represents the variables used in this study to understand their association and examine the role of AI in workforce development and provide practical insights for organizations to implement AI-driven initiatives effectively.

Table 1. Represents the factors used in the study.

Variables	Abbreviation
Skill Identification and Gap Analysis	SIGA
Employee Training Programs	ETP
Technological Integration	TI

(Table 1) cont.....

Variables	Abbreviation
Management Support and Leadership	MSL
Resource Allocation	RA
Employee Engagement and Motivation	EEM
Learning and Development Infrastructure	LDI
Organisational Culture	OC
Policy and Regulatory Support	PRS

Source: Compiled by the Authors.

ISM Methodology

- Identification of Variables: The variables under consideration are documented. The factors are taken from the review of literature and expert opinion. This tool forms the basis for analysis.
- Formation of Contextual Relationships: The connections among the identified variables are determined.
- Development of an SSIM (Self-Structured Interaction Matrix): This matrix is formed from the acknowledged variables to develop the connection among them. In a matrix, each cell indicates which factors are connected directly or indirectly.
- Construction of a Reachability Matrix: A reachability matrix is created from the ISM process, and then transitivity is checked. This background is important in ISM for identifying the levels of each variable. Transitivity ensures that if Variable A impacts Variable B and Variable B affects Variable C, then Variable A is supposed to influence Variable C indirectly.
- Partitioning Matrix: The matrix obtained in the previous step is segregated into different levels. This phase helps in organizing the hierarchical model, where the variables are situated, giving to their levels of impact. MICMAC uses this info to categorize variables into four categories: independent, dependent, linkage, and independent.
- Creation of a Graph: A diagram is drawn based on the levels known during the partition process. Another directed diagram is created to signify the associations among variables visually. It is grounded on the levels and the ordered structure determined during partitioning.
- Development of the Interpretive Model: A diagram is created to characterize the associations among variables visually. It is grounded on a classified structure determined during partitioning. In this methodology, the diagram signifies how variables are linked and their impact grading. MICMAC investigation then practices this to analyse the power of each variable. The diagram is then converted into a model by substituting variable nodes with statements.

This background uses a figure to define the nature of relationships between each pair of variables (i, j), where i) V: Variable *i* can affect variable *j* ii) A: Variable *j* affects variable *i*. iii) X: Variables *i* and *j* mutually impact each other. IV) O: Variables *i* and *j* do not stimulate each other. This process proposes identifications of how each variable moves others and helps to streamline the composite interconnections that plan the complete system. **The Structural Self-Interaction Matrix** (Table **2**) identifies the contextual relationships between the variables considered in this study using definite symbols.

Table 2. Structural Self-Interaction Matrix (SSIM).

Elements	SIGA	ETP	TI	MSL	RA	EEM	LDI	OC	PRS
SIGA		O	A	O	O	A	A	A	A
ETP			V	V	O	V	O	O	O
TI				V	V	V	V	O	V
MSL					V	A	A	A	A
RA						A	A	A	A
EEM							A	X	A
LDI								V	V
OC									X
PRS									

Source: Compiled by the Authors.

The matrix suggestions the association among key variables linked to the AI-Driven Automation for Upskilling and Reskilling the Workforce. The following relationships have been identified:

- Skill Identification and Gap Analysis (SIGA): Impacts Employee Training Programs (ETP) and Organizational Culture (OC) as the proper identification of skill gaps drives training and affects culture. SIGA is influenced by Employee Training Programs (ETP) and Technological Integration (TI), as training and technology help refine and identify necessary skills for improvement.
- Employee Training Programs (ETP): Impacts Technological Integration (TI), Resource Allocation (RA), Employee Engagement and Motivation (EEM), and Learning and Development Infrastructure (LDI). A well-structured training program ensures that employees can adapt to technological changes, ensuring effective resource allocation and engagement. Technological Integration (TI) and Learning and Development Infrastructure (LDI) have an impact on ETP since technology dictates how and what kind of training is designed and implemented.

- Technological Integration (TI): The effects of technological integration on Employee Engagement and Motivation (EEM), Organizational Culture (OC), Learning and Development Infrastructure (LDI), Resource Allocation (RA), and Management Support and Leadership (MSL). Organisational leadership, resource allocation, infrastructure development, employee engagement, and culture are all impacted by technological advancements. The seamless implementation of technology is guaranteed by Management Support and Leadership (MSL), which impacts Technology Implementation (TI).
- Management Support and Leadership (MSL): Impacts Technological Integration (TI) and Resource Allocation (RA). Leadership plays a key role in driving the adoption of technology and ensuring the effective allocation of resources. MSL is influenced by Technological Integration (TI), Resource Allocation (RA), Employee Engagement and Motivation (EEM), and Learning and Development Infrastructure (LDI).
- Resource Allocation (RA): Affects Management Support and Leadership (MSL), Employee Engagement and Motivation (EEM), Learning and Development Infrastructure (LDI), and Organizational Culture (OC). Proper allocation of resources is crucial for engaging employees and developing the necessary infrastructure. It is influenced by Management Support and Leadership (MSL), which determines how resources are distributed.
- Employee Engagement and Motivation (EEM): Influences Learning and Development Infrastructure (LDI) and Policy and Regulatory Support (PRS). Engaged and motivated employees foster a stronger learning environment and can influence policy changes.EEM is influenced by Resource Allocation (RA) and Policy and Regulatory Support (PRS), as resource distribution and policies impact engagement levels.
- Learning and Development Infrastructure (LDI): Influences Policy and Regulatory Support (PRS), as a strong learning infrastructure can push for regulatory support. LDI is influenced by Technological Integration (TI), Employee Engagement and Motivation (EEM), and Organizational Culture (OC).
- Organizational Culture (OC): Influences Policy and Regulatory Support (PRS), as the prevailing culture within an organization drives how policies are shaped and supported. It is influenced by Skill Identification and Gap Analysis (SIGA), Technological Integration (TI), and Resource Allocation (RA).
- Policy and Regulatory Support (PRS): impacted by Organizational Culture (OC), Employee Engagement and Motivation (EEM), and Learning and Development Infrastructure (LDI), as culture, employee motivation, and infrastructure collectively impact regulatory policies.

Reachability Matrix

The Reachability Matrix (Table **3**) highlights the driving and dependency powers of the nine identified variables, emphasizing their interconnectedness.

Table 3. Reachability matrix.

Variables	SIGA	ETP	TI	MSL	RA	EEM	LDI	OC	PRS	Driving Power
SIGA	1	0	0	0	0	0	0	0	0	1
ETP	0	1	1	1	0	1	0	0	0	4
TI	1	0	1	1	1	1	1	0	1	7
MSL	0	0	0	1	1	0	0	0	0	2
RA	0	0	0	0	1	0	0	0	0	1
EEM	1	0	0	1	1	1	0	1	0	5
LDI	1	0	0	1	1	1	1	1	1	7
OC	1	0	0	1	1	1	0	1	1	6
PRS	1	0	0	1	1	1	0	1	1	6
Dependency Power	6	1	2	7	7	6	2	4	4	

Source: Compiled by the Authors.

The matrix above summarises the driving power and dependency power of the nine identified variables within the study, stressing the extent to which each variable impacts and is influenced by others. Skill Identification and Gap Analysis (SIGA) has a driving power of 1, indicating it is independent with negligible influence on other variables. On the other hand, Technological Integration (TI) and Learning and Development Infrastructure (LDI) have the maximum driving power, each at 7, viewing that they apply substantial effect across various factors such as Employee Training Programs (ETP), Management Support and Leadership (MSL), Resource Allocation (RA), Employee Engagement and Motivation (EEM), Organizational Culture (OC), and Policy and Regulatory Support (PRS). These factors play a significant role in shaping workforce development. Management Support and Leadership (MSL) and Resource Allocation (RA), with a driving power of 2 and 1, respectively, reflect their more dependent parts. The Employee Engagement and Motivation (EEM) and Organizational Culture (OC) show a reasonable driving power of 5 and 6, stressing their twofold role in influencing and being influenced by other variables. Policy and Regulatory Support (PRS) likewise grips a moderate position, showing both influence and dependency. The dependence power highlights Management Support and Leadership (MSL) and Resource Allocation (RA) as vastly dependent, determined by other factors, while Skill Identification and Gap

Analysis (SIGA) remains the least dependent. This analysis highlights the interrelatedness among these variables, locating Technological Integration (TI) and Learning and Development Infrastructure (LDI) as significant drivers in the system.

Final Reachability Matrix

Refined analysis of variables, the Final Reachability Matrix (Table **4**), summarizes the hierarchical structure of the variables.

Table 4. Final Reachability Matrix

Variables	SIGA	ETP	TI	MSL	RA	EEM	LDI	OC	PRS	Driving Power
SIGA	1	0	0	0	0	0	0	0	0	1
ETP	1*	1	1	1	1*	1	1*	1*	1*	9
TI	1	0	1	1	1	1	1	1*	1	8
MSL	0	0	0	1	1	0	0	0	0	2
RA	0	0	0	0	1	0	0	0	0	1
EEM	1	0	0	1	1	0	0	1	1*	6
LDI	1	0	0	1	1	1	1	1	1	7
OC	1	0	0	1	1	1	0	1	1	6
PRS	1	0	0	1	1	1	0	1	1	6
Dependency Power	7	1	2	7	8	6	3	6	6	

Source: Compiled by the Authors.

The matrix tells the driving and dependency powers of various variables influencing AI-driven upskilling and reskilling initiatives. Employee Training Programs appear as the most crucial factor, with the highest driving power (9), signifying its key role in influencing other factors. Technological Integration follows thoroughly with a driving power of 8, emphasizing the role of integrating technology into workforce development. Factors such as Management Support and Leadership, and Resource Allocation have small driving power, demonstrating their dependency on other aspects for success. Mid-range factors, for instance, Learning and Development, Infrastructure, and Organizational Culture, play a key role in linking important elements of the framework. The Policy and Regulatory Support (PRS) and Employee Engagement and Motivation (EEM) factors demonstrate sensible roles, contributing to both driving and dependent powers, making them fundamental to the overall system. Overall, this matrix emphasis that strategic focus on training programs and technological integration will initiate the success of AI-driven workforce transformation.

Level Partitioning

The Level Partitioning Iterations in Tables **5** and **6** illustrate the classification of variables based on their reachability and dependency sets.

Table 5. Level partitioning iterations.

Elements)	Reachability Set	Antecedent Set	Intersection Set	Level
SIGA	1,	1, 2, 3, 6, 7, 8, 9,	1,	1
ETP	1, 2, 3, 4, 5, 6, 7, 8, 9,	2	2	
TI	1, 3, 4, 5, 6, 7, 8, 9,	2, 3	3	
MSL	4, 5,	2, 3, 4, 6, 7, 8, 9,	4,	
RA	5	2, 3, 4, 5, 6, 7, 8, 9,	5	1
EEM	1, 4, 5, 6, 8, 9,	2, 3, 6, 7, 8, 9	6, 8, 9	
LDI	1, 4, 5, 6, 7, 8, 9,	2, 3,7,	7,	
OC	1, 4, 5, 6, 8, 9,	2, 3, 6, 7, 8, 9	6, 8, 9,	
PRS	1, 4, 5, 6, 8, 9,	2, 3, 6, 7, 8, 9,	6, 8, 9,	
		1 2 3 4 5 6		

Source: Compiled by the Authors.

Table 6. Level partitioning.

Elements)	Reachability Set	Antecedent Set	Intersection Set	Level
SIGA	1,	1, 2, 3, 6, 7, 8, 9,	1,	1
ETP	1, 2, 3, 4, 5, 6, 7, 8, 9,	2	2	
TI	1, 3, 4, 5, 6, 7, 8, 9,	2, 3	3	
MSL	4, 5,	2, 3, 4, 6, 7, 8, 9,	4,	
RA	5	2, 3, 4, 5, 6, 7, 8, 9,	5	1
EEM	1, 4, 5, 6, 8, 9,	2, 3, 6, 7, 8, 9	6, 8, 9	
LDI	1, 4, 5, 6, 7, 8, 9,	2, 3,7,	7,	
OC	1, 4, 5, 6, 8, 9,	2, 3, 6, 7, 8, 9	6, 8, 9,	
PRS	1, 4, 5, 6, 8, 9,	2, 3, 6, 7, 8, 9,	6, 8, 9,	
		1 2 3 4 5 6		

Source: Compiled by the Authors.

The iteration (Table **5**) supports breaking down the elements into diverse levels based on their reachability, antecedent, and intersection sets. This process is decisive to considerate the classified construction of the variables.

- Skill Identification and Gap Analysis is ranked at level 1 because its reachability and intersection sets are the same (only includes itself). This means SIGA doesn't depend on other factors and is considered an independent driver.
- Employee Training Programs is ranked at level 2 because it is affected and inclined by multiple factors. Though it has more power than dependency, it is employed higher in the hierarchy.
- Technological Integration stood at the third level, having a more stable influence and reliance. It plays a dominant role, impacting other variables and being influenced by them too.
- Management Support and Leadership and Resource Allocation have more dependence on other factors and are positioned at transitional levels, suggesting that they depend on basic elements like training and technology.
- Factors like Employee Engagement and Motivation (EEM), Learning and Development Infrastructure, Organizational Culture, and Policy and Regulatory Support are more knotted and are placed at higher levels, representing that they are moulded by and contribute to the system's dynamics.

Each variable passes through different stages based on how much it affects or is impacted by others, making a clear, classified structure of relationships between the elements.

Diagraph

The relationships among the variables influencing AI-driven upskilling and reskilling initiatives are depicted in Fig. (**1**), which provides a diagram representation based on the ISM methodology.

Fig. (**1**) shows the relationship among various factors related to the variables influencing AI-driven upskilling and reskilling initiatives.

In the structural model, variables are categorized into different levels based on their influence and dependency within the system:

- Top-level Factors like Organizational Culture, Employee Engagement and Motivation, and Policy and Regulatory Support are highly reliant on other variables but have partial driving power. They are influenced by basic elements and signify the results in the system.
- Middle-level variables, such as Technological Integration and Learning and Development Infrastructure, show a balanced effect and reliance. These variables act as critical enablers in the system, both influencing toplevel factors and being squeezed by lower-level variables such as skill identification and resource allocation.

- Lower-level variables, including Skill Identification and Gap Analysis, and Resource Allocation, have solid driving power with limited reliance. These basic elements initiate and influence the system, setting the stage for consequent decisions.
- Interconnected variables like Employee Training Programs and Management Support, and Leadership work across levels, affecting both basic and higher-order factors. They fill the gap between lower and top levels, simplifying the flow of activities within the system.

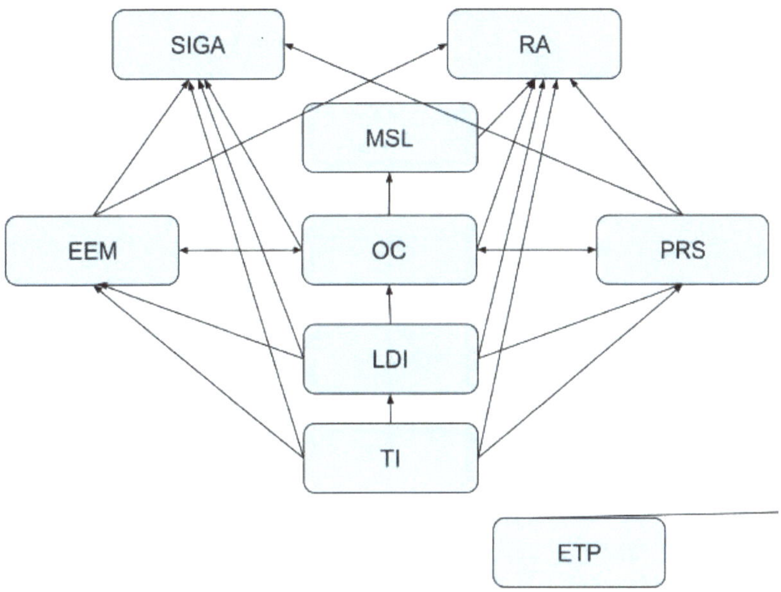

Fig. (1). A Diagram representation of the variables.

This hierarchy strengthens the relationships among the variables, classifying which factors drive the system.

A simplified representation based on the hierarchy described in ISM results is given below:

MICMAC Analysis

The MICMAC analysis in Fig. (**2**) categorizes the factors into independent, dependent, linkage, and driving variables, illustrating their respective roles in the system.

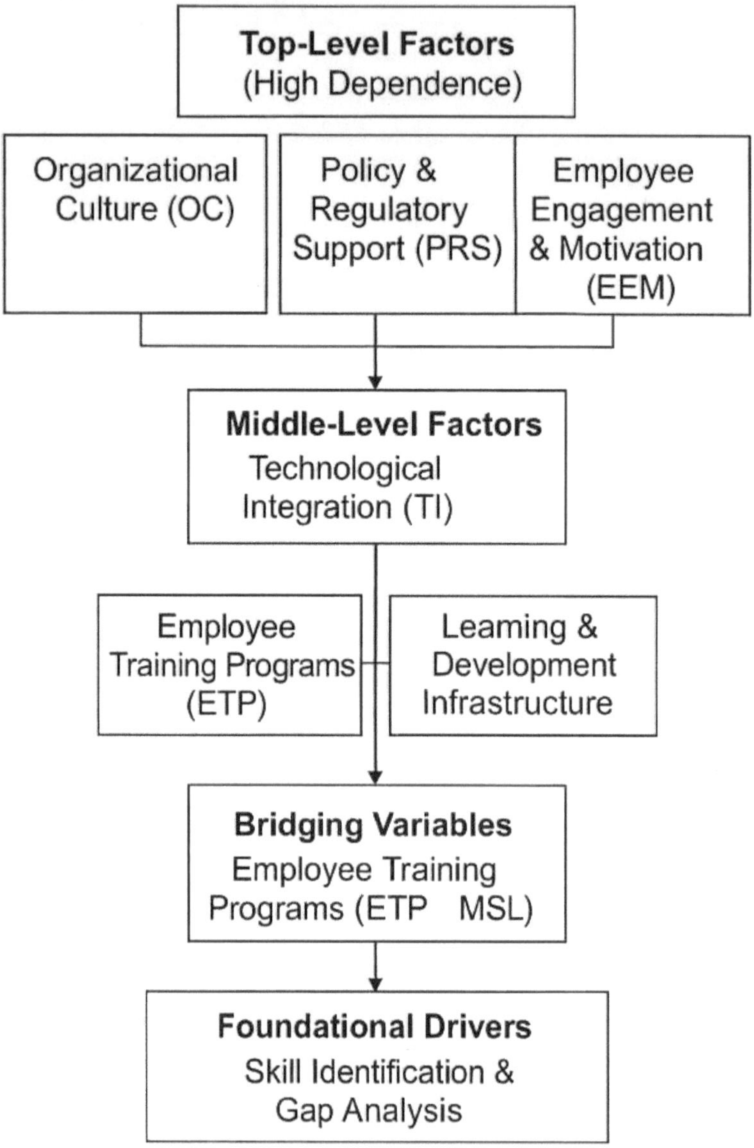

Fig. (2). ISM Hierarchy of factors influencing ai-driven reskilling.

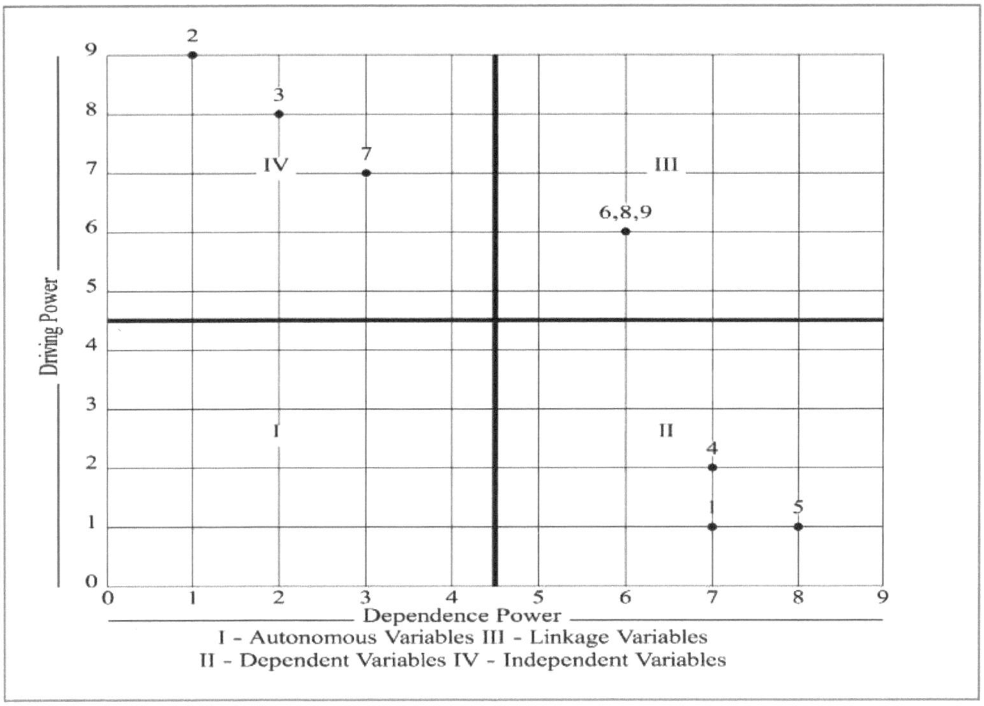

Fig. (3). Shows a cross-impact matrix multiplication applied to classification (MICMAC) analysis of AI-driven automation for upskilling and reskilling the workforce.

The MICMAC analysis categorizes the factors into four categories based on their driving and dependence power. Independent factors like Skill Identification and Gap Analysis have low driving and dependence power, stated that they are disengaged and do not pointedly influence or get influenced by others. Dependent variables, such as Policy and Regulatory Support, Organizational Culture, and Employee Engagement and Motivation, have high dependency power but low driving power, making them highly impacted by other factors but poor in shaping the system. Linkage variables, including Technological Integration, Employee Training Programs, and Learning and Development Infrastructure, grip both high driving and dependence power, acting as crucial factors that affect and are impacted by others, possibly causing significant shifts in the system. Lastly, driving variables like Resource Allocation and Management Support and Leadership have high driving but low dependence power, applying solid influence over other variables while being less affected themselves, making them the key drivers of the system. This analysis benefits from identifying key variables that require strategic focus for effective intervention and system stability.

The MICMAC analysis categorizes variables based on their driving power (how much they influence others) and dependence power (how much they are influenced).

The categories shown in Table **7** help prioritize which factors to invest in, which to monitor, and which are early warning indicators in AI-driven reskilling efforts.

Table 7. Understanding MICMAC categories.

Quadrant	Category	What it Means	Example from Study
I	**Autonomous**	Weak drivers and weak dependencies — mostly isolated	Not applicable in this study
II	**Dependent**	Strongly affected by others, but doesn't affect others much	Policy and Regulatory Support, OC, EEM
III	**Linkage**	Affect others and are affected back — unstable and critical	Tech Integration, Training Programs, LDI
IV	**Independent**	Strong influencers but not much affected — strategic levers	Resource Allocation, Management Support

DISCUSSION

This study, focusing on AI-Driven Automation for Upskilling and Reskilling the Workforce using ISM and MICMAC analysis, categorizes factors that affect successful implementation. Resource Allocation (RA) and Management Support and Leadership (MSL) are critical factors that drive the whole system. They are vital for effectively using AI in upskilling programs. Technological Integration and Employee Training Programs are examples of linkage variables that exhibit significant interconnectedness and contribute to the process by acting as both influencers and influenced factors. Regulatory support, organisational culture, and frameworks are all dependent factors that highlight the need for policies and frameworks to guarantee AI-driven change. This planned study highlights the importance of leadership and investing resources, while also emphasising the necessity for a legislative and cultural climate that supports worker upskilling in the era of AI. With the integration of the technical and ethical considerations, this research study fills the research gap and serves as a roadmap for aligning AI-driven automation with organizational development goals.

RECOMMENDATIONS AND CONCLUSIONS

The purpose of this research is to highlight essential concepts for the application of AI-driven automation for the purpose of upskilling and reskilling the workforce. To guarantee that AI projects have access to enough human, financial, and technical resources, organisations should prioritise resource allocation strategies. Leadership and support from upper management are also essential; executives should push for these projects with all their might so that they align with the company's goals. To make unified AI integration into current processes easier, technical integration should be strengthened. To promote a culture of continuous learning and solve individual skill shortages, staff training programs should be customised. When it comes down to it, the key to a successful workforce upskilling in the AI era is a supportive organisational culture that embraces change and the development of clear policy and legal frameworks. These factors will increase acceptance and involvement in AI-driven projects.

Limitations and Scope for Further Research

These findings and their generalisability may be affected by a number of limitations. Firstly, there is a lack of capacity to quantify the influence of different elements, and the research is mostly dependent on qualitative data, which might induce bias. Secondly, the effects may not be generalisable to other areas if the research only looked at one or two businesses. Also, certain results may become outdated as a result of AI technology progress; therefore, there will need to be continuous updates to the study.

Research on the long-term effects of upskilling programs powered by AI in various sectors has a lot of room to grow. For a fuller picture of the interrelationships of the recognised components, empirical investigations using quantitative data could be useful. Upskilling programs might also benefit from investigating how new technologies like data analytics and machine learning fit into the picture. Last but not least, to fully grasp the effect of AI-driven upskilling on worker performance and organisational success, longitudinal studies are needed to assess its efficacy over time.

POLICY IMPLICATIONS

This study brings significant policy implications for organisations and governments aiming to effectively implement AI-driven automation for workforce upskilling and reskilling. Firstly, policymakers should align with a supportive regulatory framework that encourages investment in AI technologies and training programs. This includes providing motivations for organisations that invest in workforce development and adopt AI solutions. Secondly, guidelines should encourage collaborations among educational institutions and industries to ensure that training programs align with market trends. This association can bridge the skills gap and prepare the workforce for future demands. Moreover, policymakers should also focus on fostering a culture of learning environment within organisations, which can be maintained through funding for professional development initiatives. Finally, ongoing evaluation and adaptation of guidelines will be important to keep pace with the technology and workforce needs, ensuring that AI-driven upskilling initiatives remain effective and relevant in the growing job market.

CONCLUSION

The conclusion highlights the strategic importance of AI-driven automation in workforce upskilling and reskilling initiatives. Resource allocation, leadership support, technological integration, and a learning-oriented organizational culture are identified as pivotal drivers for implementing effective AI-enabled workforce development programs. Employee training programs, when customized to individual needs, facilitate skill acquisition, enhancing productivity and adaptability. A supportive policy framework and collaborative efforts between organizations and educational institutions are essential to align training programs with evolving market needs. This research fills a critical gap by integrating technical and ethical dimensions to provide a comprehensive framework for workforce transformation, ensuring both organizational competitiveness and employee empowerment in the age of AI.

REFERENCES

Acemoglu, D. (1998). Why do new technologies complement skills? *Q. J. Econ., 113*(4), 1055-1089. [http://dx.doi.org/10.1162/003355398555838]

Alfes, K., Shantz, A.D., Truss, C., Soane, E.C. (2013). The link between perceived human resource management practices, engagement and employee behaviour: a moderated mediation model. *Int. J. Hum. Resour. Manage., 24*(2), 330-351. [http://dx.doi.org/10.1080/09585192.2012.679950]

Autor, D.H. (2015). Why are there still so many jobs? The history and future of workplace automation. *J. Econ. Perspect., 29*(3), 3-30. [http://dx.doi.org/10.1257/jep.29.3.3]

Berg, A., Buffie, E.F., Zanna, L.F. (2018). Should we fear the robot revolution? (The correct answer is yes). *J. Monet. Econ., 97*, 117-148. [http://dx.doi.org/10.1016/j.jmoneco.2018.05.014]

Bhardwaj, R., Kumar, V. (2022). AI-based learning systems: An innovative approach to skill acquisition. *J. Bus. Res., 138*, 184-192. [http://dx.doi.org/10.1016/j.jbusres.2021.09.013]

Brynjolfsson, E., Hitt, L.M. (1998). Beyond the productivity paradox. *Commun. ACM, 41*(8), 49-55. [http://dx.doi.org/10.1145/280324.280332]

Caldwell, M. (2015). The role of government in promoting AI adoption and workforce development. *Public Policy Res., 11*(1), 45-62. [http://dx.doi.org/10.1177/0013161X14549268]

Cameron, K.S., Quinn, R.E. (2011). *Diagnosing and changing organizational culture: Based on the competing values framework.*. Addison-Wesley.

Carvalho, A., Abraham, J. (2022). Government policies and the ethical deployment of AI in workforce training. *Technol. Soc., 68*, 101795.

[http://dx.doi.org/10.1016/j.techsoc.2022.101795]

Chauhan, A., Thakur, R., Srivastava, M. (2020). Adaptive learning environments through AI: Enhancing employee engagement and learning outcomes. *Int. J. Inf. Manage., 50*, 132-141.
[http://dx.doi.org/10.1016/j.ijinfomgt.2019.06.008]

Dissanayake, D., Ranjit, J., Gunawardena, N. (2023). The impact of AI-driven performance feedback on employee motivation: A study on reskilling initiatives. *J. Workplace Learn., 34*(3), 193-206.
[http://dx.doi.org/10.1108/JWL-07-2021-0117]

Dwivedi, Y.K., Ismagilova, E., Hughes, L., Kholod, A. (2023). Artificial intelligence (AI): Multidisciplinary perspectives on emerging challenges, opportunities, and threats. *Int. J. Inf. Manage., 57*, 102201.
[http://dx.doi.org/10.1016/j.ijinfomgt.2020.102201]

Garavan, T.N., Carbery, R., Rock, A. (2016). The role of learning and development in achieving organizational goals: A systematic literature review. *Int. J. Hum. Resour. Manage., 27*(10), 1119-1139.
[http://dx.doi.org/10.1080/09585192.2016.1201290]

Hager, P., Gonczi, A. (1996). *Key competencies: The missing link in the education and training of professionals. Australian Vocational Education and Training Research Association..* AVETRA.

Hancock, S., Dierdorff, E.C., Ruderman, M. (2020). Upskilling for the future: The role of employee development in maintaining a competitive edge. *Hum. Resour. Manage. Rev., 30*(4), 100730.
[http://dx.doi.org/10.1016/j.hrmr.2019.100730]

Jaiswal, A., O'Donnell, M., Karr, W. (2022). Multilevel HR frameworks and AI-driven workforce development: Bridging the gap. *Hum. Resour. Dev. Q., 33*(1), 15-36.
[http://dx.doi.org/10.1002/hrdq.21402]

Kaplan, J., Haenlein, M. (2019). Siri, Siri, in my hand: A critical review of the role of artificial intelligence in marketing. *Bus. Horiz., 62*(6), 825-836.
[http://dx.doi.org/10.1016/j.bushor.2019.07.006]

Kohli, R., Melville, N. (2019). Digital innovation: A framework for understanding and managing technology adoption. *J. Strateg. Inf. Syst., 28*(1), 1-20.
[http://dx.doi.org/10.1016/j.jsis.2019.03.001]

Lee, J., Lee, K. (2020). The role of AI in personalized learning: Applications in human resource management. *Int. J. Hum. Comput. Interact., 36*(8), 754-766.
[http://dx.doi.org/10.1080/10447318.2019.1645703]

Leinen, A., Diez, F., Karpov, V. (2020). Job displacement and the future of work in the age of AI: A systematic review. *Work Employ. Soc., 34*(5), 763-785.
[http://dx.doi.org/10.1177/0950017019881183]

Levy, F. (2018). Computers and populism: artificial intelligence, jobs, and politics in the near term. *Oxf. Rev. Econ. Policy, 34*(3), 393-417.
[http://dx.doi.org/10.1093/oxrep/gry004]

Levy, F., Murnane, R. (2004). The new division of labor: How computers are creating the next labor market. *Polit. Soc., 32*(3), 367-407.
[http://dx.doi.org/10.1177/0032329204267975]

Malik, A., Budhwar, P., Patel, C., Srikanth, N.R. (2022). May the bots be with you! Delivering HR cost-effectiveness and individualised employee experiences in an MNE. *Int. J. Hum. Resour. Manage., 33*(6), 1148-1178.
[http://dx.doi.org/10.1080/09585192.2020.1859582]

Malik, M., Moin, M., Khan, S. (2022). The impact of AI on HRM in the post-COVID-19 era: Remote and hybrid work. *Int. J. Hum. Resour. Manage., 33*(2), 329-354.
[http://dx.doi.org/10.1080/09585192.2021.1973273]

Mohri, M., Rostamizadeh, A., Talwalkar, A. (2018). *Foundations of machine learning.* MIT Press.

Nguyen, T. M., & Malik, A. (2021). A two-wave cross-lagged study on AI service quality: The moderating effects of the job level and job role. British Journal of Management, 1–17 (online). [http://dx.doi.org/10.1111/1467-8551.12540]

Puzzo, A., Larese, T., Musmeci, A. (2023). New job categories and the evolving labor market in the AI era. *J. Labor Res., 41*(1), 64-83. [http://dx.doi.org/10.1007/s12122-020-09342-5]

Rahman, A., Abedin, M.J. (2021). The Fourth Industrial Revolution and private commercial banks: the good, bad and ugly. *Int. J. Organ. Anal., 29*(5), 1287-1301. [http://dx.doi.org/10.1108/IJOA-05-2020-2218]

Rao, P., Prasad, S. (2022). Enhancing employee performance through technological adoption: A study on AI tools. *J. Bus. Res., 139*, 367-378. [http://dx.doi.org/10.1016/j.jbusres.2021.10.036]

Russell, S., Norvig, P. (1995). *Artificial intelligence: A modern approach.*. Prentice Hall.

Saukkonen, M., Rajala, R., Kuusisto, A. (2019). The role of AI in transforming HRM practices: Opportunities and challenges. *J. Bus. Res., 98*, 201-210. [http://dx.doi.org/10.1016/j.jbusres.2019.01.045]

Sharma, A., Mishra, R. (2019). Resource constraints and AI adoption in workforce training. *Journal of Strategic Management, 12*(3), 221-234. [http://dx.doi.org/10.1016/j.jsm.2019.07.002]

Silva, A., Santos, F., Marques, J. (2021). Organizational culture and AI adoption: Impact on employee participation in upskilling programs. *J. Organ. Change Manage., 34*(6), 1321-1339. [http://dx.doi.org/10.1108/JOCM-06-2021-0212]

Singh, A., Thakur, R., Gupta, S. (2021). Investing in AI technologies for workforce productivity: An empirical analysis. *Int. J. Prod. Perform. Manag., 70*(3), 540-556. [http://dx.doi.org/10.1108/IJPPM-01-2020-0027]

Stoker, J. (2019). Transformational leadership and employee engagement in the context of AI: Implications for HRM. *Int. J. Hum. Resour. Manage., 30*(8), 1261-1280. [http://dx.doi.org/10.1080/09585192.2018.1427636]

Tommasi, F., Perini, M., Sartori, R. (2022). Multilevel comprehension for labor market inclusion: a qualitative study on experts' perspectives on Industry 4.0 competences. *Educ. Train., 64*(2), 177-189. [http://dx.doi.org/10.1108/ET-04-2021-0146]

Vrontis, D., Thrassou, A., Lamprianou, I. (2021). The influence of AI on human resource management: A systematic literature review. *Int. J. Hum. Resour. Manage., 32*(8), 1781-1801. [http://dx.doi.org/10.1080/09585192.2020.1772085]

World Economic Forum. (2020). The Future of Jobs Report 2020. Available from: https://www.weforum.org/reports/the-future-of-jobs-report-2020.

CHAPTER 7

The Skills Revolution for the Age of AI: Preparing for the Future

Sarita Rana[1,*] and **Supriya Pathak**[2]

[1] *Acropolis Faculty of Management and Research, Indore, India*

[2] *Faculty of Management, Oriental University, Indore, India*

Abstract: A skills revolution is required due to the emergence of AI, focusing on reskilling and upskilling to meet new demands. This paper examines the need for digital literacy, advanced cognitive skills, and ethical issues. We can develop a resilient labor force skilled at navigating an AI-driven future by encouraging lifelong learning and cooperation. The urgent necessity for reskilling and up-skilling to meet the needs of an AI-driven economy is examined in this study. This paper emphasizes a proactive approach to workforce development in the era of AI, highlighting key projects and tactics for navigating the skills revolution. The ultimate objective is to establish a robust and equitable economy where AI and people collaborate to promote wealth and advancement.

Keywords: Artificial Intelligence (AI), Cognitive abilities, Digital literacy, Decision-making, Emotional intelligence, Ethical issues, Lifelong learning, Problem-solving, Training programs, Workforce transformation.

INTRODUCTION

Artificial Intelligence (AI) is transforming several sectors and changing the nature of the workforce. The demand for new skill sets prioritizing sophisticated cognitive ability, digital literacy, and emotional intelligence is rising as AI technologies automate ordinary activities. This change necessitates a thorough strategy to upskill people to ensure they are prepared to prosper in an AI-driven society. Corporations, governments, and educational institutions must work together to promote flexibility and lifelong learning. This study looks at the tactics needed to deal with this skills revolution. It emphasizes how crucial it will be to have a flexible, morally conscious, and technologically savvy workforce in the future.

* **Corresponding author Sarita Rana:** Acropolis Faculty of Management and Research, Indore, India; E-mail: dr.ranasaritag@gmail.com

Rishikaysh Kaakandikar, Keshav Kaushik, Supriya Pathakm Madhavi Shamkuwar & Vivek Mishra (Eds.)

The emergence of Artificial Intelligence (AI) is changing businesses and altering the workforce as we approach an unparalleled technological era. This revolutionary wave necessitates a significant change in the skill set and is not just an evolution. AI technologies are developing at a rapid pace, which brings with it both potential and difficulties. As a result, to stay relevant and competitive, individuals and organizations must adapt quickly.

AI automates repetitive jobs, allows sophisticated problem-solving capabilities, and revolutionizes work. The relevance of digital literacy, critical thinking, and adaptability must be emphasized in reevaluating conventional skill sets that result from this transformation. Human jobs are changing to become more strategic, creative, and interpersonal as robots take over monotonous and data-intensive work. As a result, there's a rising need for AI-complementing talents like emotional intelligence, creativity, and sophisticated decision-making.

Systems of education and training must adapt to meet these shifting demands. To stay up to speed with technology breakthroughs, people now need to continuously refresh their abilities, making lifelong learning a need rather than a luxury. This covers both soft skills that improve human-AI collaboration and technical abilities linked to AI and machine learning. Governments, universities, and corporations need to work together to build strong frameworks that encourage lifelong learning and skill improvement. Furthermore, the introduction of AI into the workforce raises moral questions and calls for laws guaranteeing fair access to jobs and education. To address these issues, a multifaceted strategy that strikes a balance between social well-being and technical advancement is needed.

Success in this new era will mostly depend on one's capacity for adaptation and learning. If we embrace the skills revolution and are ready for it, we can fully utilize AI's potential to build a more inventive, effective, and inclusive society. The first steps on the trip involve reevaluating education, promoting a culture of ongoing development, and ensuring that everyone has the chance to prosper in the AI era.

Artificial Intelligence (AI) advancements are rapidly transforming various industries, profoundly impacting workforce development and employee roles. As AI adoption accelerates, multinational corporations focus on upskilling employees to meet the demands of this evolving landscape (Jaiswal *et al.*, 2022). The introduction of generative AI shapes new trends and opportunities, offering transformative potential across multiple sectors (Jovanovic and Campbell, 2022). Within the healthcare sector, AI is viewed as a utopia and dystopia, depending on its impact on clinical roles and skills, with experts highlighting the need for careful integration (Aquino *et al.*, 2023). Similarly, intelligent automation reshapes knowledge and service work, necessitating strategic planning to capitalize on its benefits fully (Coombs *et al.*, 2020). As businesses embrace automation, understanding its fundamentals and strategic application is crucial for optimizing workplace efficiency and productivity (Chui *et al.*, 2015). Moreover, the theory of technology-mediated learning is gaining attention for its relevance in adapting to these changes in the educational and professional domains (Bower, 2019). The ongoing acceleration of technological change, especially in AI, is thus reshaping our future, making it vital for organizations to stay ahead by continuously investing in workforce development and automation technologies (Butler, 2016; Kunnen&Bosma, 2003).

Objectives

- To examine the superior cognitive abilities and vital digital literacy required for an AI-driven economy.
- Promote ongoing education and skill development to keep the workforce flexible and competitive.
- Analyze the ethical issues artificial intelligence raises and provide ways to ensure technology is used fairly and responsibly.
- Develop plans to provide workers with the skills to survive and navigate an AI-driven world.
- Encourage collaboration between government agencies, corporations, and educational institutions to promote workforce development.
- Highlight important programs and strategies for preparing the workforce for the needs of the AI era.

Literature Review

The area of Artificial Intelligence (AI) in the workplace and business has seen several recent advancements. In the past, AI-based systems used Natural Language Processing (NLP) and Artificial Intelligence (AI) to comprehend and simulate human interaction with computer systems, automating various back-office tasks, including data input, document management, accounting, and customer support (Jaiswal and colleagues, 2022; Butler, 2016).

Campbell and Jovanovic's (2022) "Generative AI" has significantly changed the game. Systems that create new material or data instead of merely processing or analyzing preexisting data are called generative AI. After learning from a piece of data, these systems can produce new information that is stylistically or semantically comparable to the original data.

Additionally, generative AI systems replicate or replace human transversal abilities like communication, problem-solving, and conflict resolution. For instance, an AI system with Natural Language Processing (NLP) skills may comprehend customer discussions, decipher their feelings, and respond in a kind and helpful manner (Jaiswal and others, 2022).

Aquino and colleagues (2023) have clarified concerns regarding the possibility of healthcare personnel losing competency due to AI advancements. After evaluating qualitative and semi-structured interviews with seventy-two healthcare professionals with AI expertise, the authors discovered two divergent opinions about the possible effects of AI on clinical abilities, such as reasoning in diagnostic and screening processes. The "utopian view" held that AI may enhance clinical expertise and impact workers' skill sets.

AI systems are employed in retail for inventory management, customer service, and sales; manufacturing for quality control or assembly; and finance departments for financial data analysis or report generation. Employees who have completed these jobs may lose their abilities due to AI, as they are no longer required to do so (Coombs *et al.*, 2020).

A number of ideas and methods explain the influence of Artificial Intelligence (AI) on human abilities in companies. One of the most widely accepted theories predicts that the advent of AI will result in a positive transformation, whereby human capacities will be improved, leading to greater efficiency and production. This theory is based on technology-mediated learning (Gupta & Bostrom, 2009; Bower, 2019).

An additional illustration may be found in a paper that explores how AI can help with healthcare labor shortages by enabling Big Data analytics, administration, diagnosis, and decision-making. The authors contend that human interaction, empathy, and effective communication are still crucial components of the care process and that AI cannot replace them. No software, application, or gadget can replace personal relationships and trust. In other words, although AI has the potential to be an invaluable cognitive aid, the job of the human doctor is unavoidable. The authors contend that while AI is not intended to take the position of healthcare professionals, those who employ it stand to gain a competitive edge over those who are ignorant of its potential applications (Meskó and associates, 2018).

Chui and colleagues (2015) discovered that AI systems could automate a sizable portion of the operations carried out by people in high-paying occupations, including executives, doctors, and portfolio managers. The majority of business occupations entail mechanical duties (like scheduling and taking attendance), cognitive tasks (like assessing client preferences and organizing logistics), and feeling tasks (like advising patients and empathizing with customers).

Workers should be able to recognize the precise knowledge and abilities required to integrate AI successfully into their work, then dissect current skills and pick up new ones to stay competitive and employable, even as AI experts relentlessly create and train machine learning algorithms to mimic human abilities. To adjust to shifting market conditions and seize new possibilities, it can be necessary to continuously educate, retrain, re-learn, or acquire new skills. According to the dynamic skills theory, skill development is a network of outcome-oriented and context-specific actions (Bosma and Kunnen, 2003).

Needs of Reskilling and Upskilling for the age of AI

Reskilling is teaching staff members new abilities so they may transition into various positions within the company. This strategy is essential when technology developments render some employment obsolete. For example, reskilling programs can assist workers in moving into professions requiring more sophisticated problem-solving and strategic thinking when AI systems take over ordinary activities.

However, upskilling aims to increase an employee's present skill set so that they may perform better in their current jobs. When the demands of their jobs change, this kind of training guarantees that workers will still be capable and productive. For instance, it becomes more critical to upskill marketing professionals to use new AI technologies efficiently when incorporated into marketing plans.

Instances within the Industry

AI is having a revolutionary effect on all industries, which calls for concentrated reskilling and upskilling efforts:

Manufacturing

Retraining programs for assembly line personnel to handle sophisticated manufacturing processes or to transfer into maintenance and troubleshooting jobs are crucial as AI-driven robotics and automation grow more common.

Healthcare

As AI is adopted in diagnostics and patient care, healthcare personnel must upskill in areas like data analysis and AI-based diagnostic tools to ensure they can use these technologies to enhance patient outcomes.

Finance

Financial organizations increasingly use AI to detect fraud and analyse risk. Programs to retrain traditional bank staff in AI and cybersecurity are essential for adjusting to these developments. An overview of key AI technologies impacting workforce transformation is presented in Table **1**.

Table 1. Overview of key AI technologies impacting workforce transformation.

Response Option	Percentage	Number of Responses (out of 58)
Very Important	50.0%	29
Important	36.2%	21
Neutral	13.8%	8
Not Important	0.0%	0

According to the respondents, reskilling is crucial in the era of artificial intelligence (AI), as seen in Table **1**. Reskilling was regarded as "Very Important" by a considerable proportion of the 58 respondents, or 50.0% (29 respondents). This suggests that 50% of participants firmly believe that upskilling is essential to adjusting to AI-driven changes in the workplace. Additionally, 36.2% of respondents (21 people) said that reskilling was "Important," indicating a significant recognition of its importance, even if it is not seen as urgent. A lesser percentage of respondents (8, or 13.8%) took a "Neutral" position, indicating a lack of certainty or a balanced opinion regarding the need for reskilling. Interestingly, not a single responder chose "Not Important," indicating that everyone agreed on the importance and necessity of developing AI-related skills. According to the data, there appears to be broad agreement among respondents regarding the significance of providing the workforce with modern skills and knowledge in response to the quick changes in technology. This emphasizes how people are becoming more conscious of how work is changing and how important it is to keep learning in order to succeed as an individual and as an organization in the AI era.

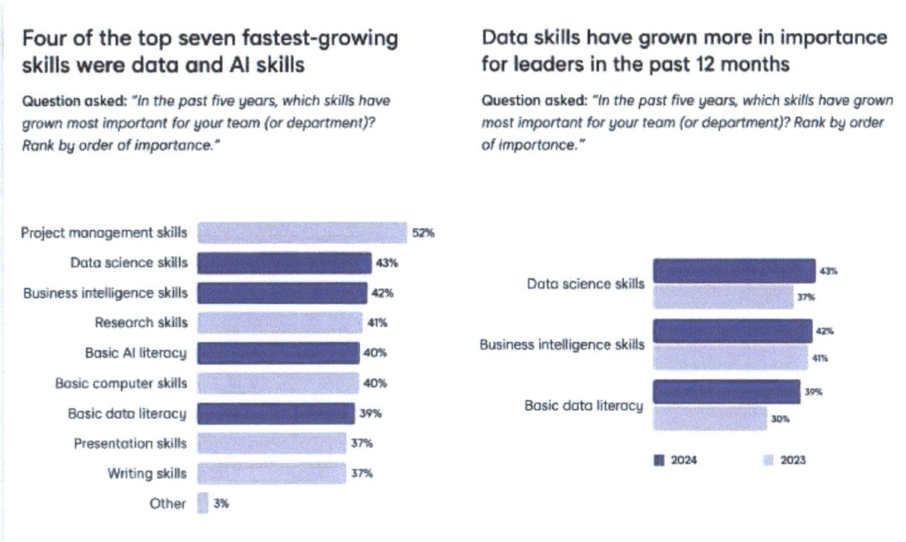

Fig. (1). The data camp state of data and ai literacy report 2024.

Source: Data from the State of Data and AI Literacy Report 2024

Explanation of the diagram:

Fastest-Growing Skills (Last 5 Years) in the Left Panel

Over the previous five years, the top abilities that have become more important in teams or departments are highlighted in the left panel. Notably, data and Artificial Intelligence (AI) account for four of the top seven skills, demonstrating the increasing importance of technological and analytical aptitude in contemporary organizations. The figures show the percentage of respondents who thought each ability was becoming more and more crucial: The growing need for data-driven decision-making is highlighted by the close follow-up of business intelligence capabilities (42%) and data science skills (43%). Research (41%), fundamental computer skills (40%), basic AI literacy (40%), and basic data literacy (37%) are other critical competencies. Organizational and leadership skills are still crucial, as evidenced by the fact that project management skills (52%) topped the list. Along with fundamental tech abilities and soft skills like writing and presentation, this trend shows a move toward digital and analytical proficiencies.

Right Panel: Data Skills' Changing Significance (2023 *vs.* 2024)

In particular, the right panel highlights how data abilities have become more crucial for leaders in the last 12 months:

The percentage of people with data science abilities rose from 37% in 2023 to 43% in 2024, indicating a notable increase in their perceived worth. Skills related to business intelligence also increased, rising from 40% to 42%.

Some important points on the importance of AI are discussed further:

AI's Transformative Potential

Artificial Intelligence (AI) is transforming the workforce by allowing more sophisticated problem-solving capabilities and automating repetitive chores. According to a McKinsey Global Institute analysis, advances in automation and AI may push 375 million people, or 14% of the world's workforce, to change occupational categories by 2030. This change calls for a thorough strategy for workforce development, emphasizing upskilling and reskilling to satisfy new requirements.

The Value of Computer Literacy

Digital literacy is the fundamental ability needed to navigate the AI-driven future. It entails an aptitude for using digital tools, comprehending data, and efficiently utilizing technology. Workers need to be adept in digital literacy to be competitive as AI becomes increasingly integrated into many sectors. According to the World Economic Forum, automation and artificial intelligence will require reskilling for 50% of all workers by 2025. In addition to increasing productivity, digital literacy allows employees to adjust to new procedures and technology.

Higher Order Cognitive Abilities

The need for sophisticated cognitive skills is growing as AI automates ordinary work. These abilities include creativity, problem-solving, and critical thinking. Artificial Intelligence (AI) systems are excellent at analyzing data and seeing patterns, but lack human creativity and nuanced comprehension. The World Economic Forum's Future of Jobs Report lists creativity, critical thinking, and complicated problem-solving as three of the most important talents that future workers will require. These kinds of workers can spur innovation and enhance AI technology.

The intelligence of emotions

In the AI age, Emotional Intelligence (EI) is yet another essential competency. It entails comprehending and controlling emotions, speaking clearly, and forming enduring bonds with others. As AI systems take on increasingly technical and data-driven activities, human jobs will shift to include more leadership, empathy, and negotiation. According to a Capgemini survey, 83% of CEOs believe that Emotional Intelligence (EI) would be a critical ability for the future workforce, indicating that EI is becoming increasingly useful as automation rises.

Lifelong Learning's Role

In the era of artificial intelligence, when continued skill improvement is necessary due to constant technological breakthroughs, lifelong learning is crucial. The fast changes wrought by artificial intelligence cannot be kept up with using traditional educational paradigms. Instead, it is necessary to promote a culture of continual learning in which people refresh their knowledge and abilities regularly. This method empowers employees to adjust to changes in the future and meet the demands of the job now. The Organisation for Economic Cooperation and Development (OECD) states that ongoing professional development and adult education are essential for ensuring economic resilience in a society driven by artificial intelligence.

Innovation and Economic Resilience

Maintaining economic resilience and promoting innovation requires funding for reskilling and upskilling programs. Task automation by AI creates new avenues for expansion and advancement. Employees with the necessary abilities can move into new positions, boosting output and the economy. According to the McKinsey Global Institute, artificial intelligence might stimulate the economy by $13 trillion by 2030.

Strategies for Enhancing and Regaining Expertise

Comprehensive upskilling and reskilling initiatives must be implemented to meet the demands of the AI age. Among these tactics are:

Government Initiatives

Governments play a critical role in advancing workforce development by sponsoring reskilling programs, providing incentives for ongoing education, and establishing laws that facilitate job transfers. For instance, Singapore's Skills Future program encourages lifelong learning and skill development by giving residents credits to undertake training programs.

Corporate Training Programs

Businesses need to spend money on training initiatives that provide staff members with the knowledge to adjust to changes brought about by artificial intelligence. Prominent corporations such as IBM and Google provide in-depth training initiatives to assist staff members in cultivating superior cognitive abilities, emotional intelligence, and digital literacy.

Public-Private Partnerships

Governments, universities, and corporations must collaborate to develop successful reskilling and upskilling initiatives. By pooling resources and knowledge, public-private partnerships may create training programs that satisfy the demands of both businesses and employees. For example, the European Union's Digital Skills and Jobs Coalition brings together companies, governments, and education providers to solve the digital skills gap.

Online Learning Platforms

The growth of online learning platforms offers flexible and easily accessible ways to enhance one's skills. People may study at their own pace and convenience with courses on AI, digital literacy, and other related skills available on platforms like Coursera, edX, and Udacity.

Industry-Specific Training

Training programs specifically designed to meet the demands of various industries are essential. For instance, the industrial sector needs expertise in AI-driven automation and robotics, while the healthcare sector needs training in AI applications for medical diagnostics.

Case Studies and Triumphant Narratives

Several case studies demonstrate how reskilling and upskilling efforts may be successfully implemented:

AT&T

In response to the swift technological advancements, AT&T initiated the extensive reskilling initiative, Future Ready. The program gives staff members access to tools for professional advancement, tuition help, and online courses. Consequently, thousands of AT&T workers have been effectively retrained, easing their move into new organizational positions.

Amazon

plans to invest $700 million to train 100,000 employees as part of its Upskilling 2025 effort. The program enables workers to advance into higher-skilled positions in machine learning and cloud computing through technical courses, apprenticeships, and on-the-job training.

Singapore Skills Future

The country of Singapore promotes lifelong learning and skill development through the Skills Future project. The program offers a variety of industry-relevant programs and grants people credits to attend training courses. Through Skills Future, Singaporeans' employability has been effectively increased, and the nation's economic growth has been promoted.

The workforce has to be reskilled and upskilled, as the AI era demands a skills revolution. By prioritizing digital literacy, sophisticated cognitive abilities, and emotional intelligence, we can cultivate a workforce that is adaptable, ethically aware, and technologically proficient. Governments, corporations, and academic institutions must work together to create comprehensive and successful training programs, and lifelong learning is a must. In light of AI's ongoing industry transformation, effective workforce development measures will guarantee economic stability, foster.

PROACTIVE APPROACHES TO ENHANCE WORKFORCE SKILLS

Proactive skill development is becoming increasingly important as Artificial Intelligence (AI) reshapes the modern workforce. Rapid technological breakthroughs in the AI age necessitate a workforce with both technical and soft abilities to handle new tasks and responsibilities. This chapter explores the proactive measures required to improve workforce competencies, highlighting the significance of lifelong learning, cooperation between firms and educational institutions, and the participation of all workers in the AI-driven economy. We can guarantee a seamless shift to AI-enhanced roles, support job stability, and encourage ongoing innovation by educating the workforce in advance.

Forecasting Skill Requirements

Projecting future skill requirements precisely is the first step towards proactively improving the workforce. This necessitates thoroughly comprehending how automation and artificial intelligence will affect many sectors and occupations. Organizations and policymakers must engage in foresight activities by utilizing data and trends to predict the talents that will be in demand. Changing how people and machines divide labor might result in the displacement of 85 million jobs by 2025. However, it could also lead to 97 million new positions better suited to the latest labor arrangement between humans, machines, and algorithms. This is according to the World Economic Forum's Future of Jobs Report.

Putting Money Into Soft and Technical Skills

It is crucial to invest in training programs that cover technical and soft skills to prepare for the future. Proficiency in technical abilities, including programming, data analysis, and AI literacy, is crucial for workers to engage with cutting-edge technology effectively. For example, understanding machine learning techniques, neural networks, and data science principles can enable employees to utilize artificial intelligence products effectively.

Technical proficiency alone, though, is insufficient. Soft talents are just as vital as complex abilities. These include creativity, problem-solving, critical thinking, and emotional intelligence. These competencies empower employees to create, cooperate efficiently, and adjust to changing work settings. Higher-order cognitive, social, emotional, and technical abilities will be in high demand by 2030; businesses such as Google and IBM already recognize the value of having a well-rounded skill set and providing training programs that include interpersonal and technical abilities.

Promoting Lifelong Education

Lifelong learning becomes essential to workforce development in the era of artificial intelligence. Traditional education models' emphasis on early-life learning and the presumption of static professions make them inadequate today. Instead, talent improvement must be ongoing throughout a person's career.

Providing chances for people to update their knowledge and skills continuously is a key component of lifelong learning. Numerous methods can accomplish this, including online courses, seminars, certificates, and on-the-job training. For instance, various cutting-edge technology and skills courses are available on platforms like Coursera, edX, and Udacity, enabling workers to learn at their own pace.

Employers are essential in encouraging lifelong learning. Businesses that provide learning incentives, such as paid time off for education, access to learning materials, and tuition reimbursement, can foster a culture of continuous development. One prominent example is AT&T's Future Ready program, which gives staff members the chance to enroll in online classes and obtain certifications in fields with high demand, such as cybersecurity and data analytics.

Partnerships Between Companies and Educational Institutions

A proactive strategy for workforce development requires close cooperation between businesses and educational institutions. This collaboration is essential to ensure that the curriculum is current and in line with industry demands. Educational institutions must update their curricula to incorporate the most recent technical developments and new skills needed in the business.

Businesses may offer insightful information about the capabilities that companies require, allowing educational institutions to modify their curricula appropriately. Co-ops, industrial collaborations, and internships can help close the knowledge gap between academic study and practical application. For instance, Northeastern University's co-op program collaborates with more than 3,000 businesses globally, allowing students to obtain real-world experience and build transferable skills they may use in their future employment.

Accessibility and Inclusion

A proactive approach to workforce development must include inclusion. It is imperative to ensure that all workers, irrespective of their background, have access to the necessary tools and resources to prosper in an economy boosted by artificial intelligence. This entails removing obstacles to education and training, such as monetary limitations, geographical restrictions, and societal prejudices.

There are several ways to implement inclusionary programs. Underprivileged groups may find it easier to acquire education and training through scholarships, grants, and financial assistance. People living in rural places can use online learning platforms, which provide them with the same chances as those living in metropolitan areas. Additionally, programs encouraging diversity in STEM disciplines can help distribute the advantages of automation and artificial intelligence fairly.

The European Union's Digital Skills and Jobs Coalition brings together businesses, governments, and educational institutions to close the digital skills gap and advance inclusiveness. This coalition is an excellent example of an inclusive strategy. The partnership wants to develop a diverse and competent workforce prepared for the AI future by concentrating on underrepresented groups, including women, older workers, and persons with impairments.

CASE STUDIES ON SELF-SUSTAINED SKILL IMPROVEMENT

In the era of artificial intelligence, several areas and organizations have successfully used proactive measures to improve workforce capabilities. These case studies offer insightful information on practical strategies and industry best practices.

The Skills Future Initiative in Singapore

Singapore's Skills Future project represents a proactive approach to workforce development. Introduced in 2015, Skills Future's goal is to encourage skill development and lifelong learning among Singaporeans. The initiative encourages a culture of lifelong learning, offers credits for skills-related courses, and promotes career counseling. Singapore invests in its workers to ensure that its people are equipped to handle the potential and difficulties of living in an AI-driven economy.

Amazon's 2025 Upskilling Plan

Through its Upskilling 2025 project, Amazon has pledged to invest $700 million to upskill 100,000 employees by 2025. The program provides a range of training options, such as certification classes, machine learning courses, and technical apprenticeships. Amazon wants to ensure its workers can fulfill high-demand tasks within the firm by proactively upskilling them and prospering in an AI-aided setting.

The Dual Education System in Germany

The dual education system in Germany serves as a template for fusing business and education. Through a combination of classroom education and on-the-job training in businesses, the program gives students real-world experience and transferable skills. Employers require graduates to have the abilities listed below to be prepared for the workforce. Thanks to the dual education system, Germany's robust manufacturing industry has been sustained, and its technical improvements have been mainly accommodated.

OVERCOMING ADOPTION CHALLENGES WITH AI

Proactive skill development is essential, but it is not without difficulties. Adopting AI and associated technology presents several challenges for both individuals and organizations. It is imperative to tackle these obstacles to facilitate a seamless shift and optimize AI's advantages.

Opposition to Change

Resistance to change is one of the main obstacles. Employees may be wary about embracing new technology and worry about losing their jobs. To overcome this, organizations need to encourage innovation and communicate the advantages of AI. One strategy to reduce resistance is to include staff members in the adoption process and offer them support and training.

Deficits in Competencies

Due to the rapid speed of technological development, the present workforce may lack the skills necessary for new occupations. Specialized training programs and collaborations with academic institutions to provide pertinent curricula are needed to address this. Companies can also offer possibilities for reskilling and upskilling to close these gaps.

Moral Issues

The adoption of AI creates ethical questions around employment displacement, prejudice, and privacy. Organizations using AI must set explicit policies and moral benchmarks. Prioritizing transparency, accountability, and justice may help foster confidence among the public and workers. Proactively addressing ethical issues can help ensure that AI is used responsibly.

Facilities and Availability

Implementing AI may be hampered by inadequate infrastructure and restricted access to technology. Governments and organizations must invest in digital tools and high-speed internet infrastructure to enable AI projects. Guaranteeing fair access to technology is equally crucial to prevent the digital gap from growing.

In Conclusion, proactively improving worker skills is crucial in the AI era to facilitate a seamless shift into new positions, provide job security, and foster creativity. This calls for a diverse strategy that involves foreseeing future skill requirements, investing in hard and soft skills, promoting lifelong learning, and boosting cooperation between businesses and educational institutions. Prioritizing accessibility and inclusion is also necessary to guarantee that all workers may prosper in an AI-enhanced economy. We can create a flexible, resilient workforce that is ready for the future by tackling these issues and taking proactive measures. Proactive skill development is beneficial, as shown by Germany's dual education system, Singapore's Skills Future initiative, and Amazon's Upskilling 2025 program. It is critical to adopt these strategies and ensure that AI's advantages are pretty and widely distributed as AI continues to transform the workforce.

WORKERS AND AI IN THE WORKPLACE

Artificial Intelligence (AI) technologies are developing so quickly that they drastically change the modern workplace. Artificial Intelligence (AI) changes employment responsibilities and workforce dynamics, from automating repetitive jobs to enhancing human talents. As Artificial Intelligence (AI) permeates more aspects of our everyday existence, the moral and societal concerns around its use in the workplace become more crucial. This chapter examines the complex effects of AI on labor transformation, job automation, and augmentation, as well as the moral and societal ramifications of these developments. Organizations and legislators may guarantee that AI technologies are created and applied to advance justice, fairness, and the welfare of all people by proactively addressing these concerns.

These results show that there is broad agreement regarding AI's revolutionary potential to redefine job duties in a variety of industries. Most respondents agree that AI improves human capabilities and opens the door for more strategic and creative roles by automating monotonous and repetitive jobs. Because of these two effects—automation and augmentation—workers must acquire new skills, adjust to changing job requirements, and pursue ongoing education. The lack of "Not Important" answers emphasizes even more how urgent it is that businesses and legislators fund AI-readiness programs, such as inclusive deployment plans, ethical frameworks, and training.

In conclusion, Table **2**'s data demonstrates the increasing understanding that integrating AI is crucial for the sustainability of the workforce in the future and is not an alternative. A fair and inclusive transition to AI-augmented workplaces requires a focus on adaptation, reskilling, and ethical AI practices.

Table 2. Automation and job augmentation: Impacts of AI on workforce roles.

Response Option	Percentage	Number of Responses (Out of 58)
Very Essential	37.9%	22
Essential	50.0%	29
Neutral	12.1%	7
Not Important	0.0%	0

Automation of Work

The use of Artificial Intelligence (AI) and other cutting-edge technology to carry out jobs previously completed by people is known as job automation. This automation may result in more productivity, lower expenses, and better accuracy in many businesses. It also prompts questions about the future of employment and job displacement. Logistics, retail, and manufacturing are just a few industries that have already seen substantial automation. For instance, the employment of robots and AI-powered devices is growing in quality control, inventory management, and assembly line labor.

AI integration in various fields has the potential to increase output and decrease human error. It also implies that many repetitive and ordinary professions risk becoming out of style. According to the Brookings Institution, approximately 25% of American occupations are very vulnerable to automation. This covers secretarial, customer service, and manufacturing jobs that need repetitive physical labor and data processing. Automation might result in job displacement, but it can also allow people to move into new tasks that require more advanced skills.

Employment Extension

Job augmentation, as opposed to automation, is the application of AI to augment rather than replace human capabilities. Artificial Intelligence (AI) can increase worker productivity and job satisfaction by assisting employees in completing tasks more successfully and efficiently.

The application of AI-powered technologies in healthcare is one instance of job augmentation. Artificial Intelligence (AI) can evaluate enormous volumes of medical data to help physicians diagnose illnesses, make treatment recommendations, and predict patient outcomes. This allows medical personnel to treat patients better and make more informed judgments.

Artificial Intelligence (AI) may help analysts with risk management, opportunity identification, and trend evaluation in the financial industry. AI-powered algorithms may process large datasets far more quickly than people, yielding insightful information that can guide strategic decision-making.

Job augmentation is also applicable in domains like customer service, where AI chatbots and virtual assistants can manage repetitive inquiries and free up human agents to work on more complicated problems. AI may help employees become more productive and efficient by enhancing human talents, eventually improving corporate outcomes and job happiness.

Transformation of the Workforce

The workforce is undergoing a dramatic transition due to the incorporation of AI. This metamorphosis includes adjustments to organizational structures, skill needs, and job responsibilities. Table **3** illustrates the shift in job roles as AI automates repetitive tasks, emphasizing higher-order cognitive and creative skills such as problem-solving and emotional intelligence. This highlights the need for continuous learning and adaptability in the workforce.

Table 3. Workforce transformation: Evolving job roles in the AI era.

Response Option	Percentage	Number of Responses (out of 58)
Very Necessary	48.3%	28
Necessary	43.1%	25
Neutral	8.6%	5
Not Necessary	0.0%	0

It is evident from the comments that there is broad agreement regarding the pressing necessity of realigning workforce roles in the AI era. Organizations must refocus their efforts on positions requiring higher-order cognitive skills like creativity, critical thinking, problem-solving, and emotional intelligence as AI replaces repetitive and regular tasks. To stay competitive and prepared for the future, this shift calls for updating job descriptions, retraining staff, and redesigning organizational structures. The idea that adjusting to AI is a strategic need rather than just an option is further supported by the total lack of "Not Necessary" answers. To effectively navigate the opportunities and difficulties brought forth by AI-driven change, the workforce must constantly develop through agility, a growth mindset, and lifelong learning. The key importance of ongoing skill development and organizational flexibility in adopting the AI-driven workforce transformation is highlighted in Table **3**.

Shifting Positions

As AI technologies enhance human skills and automate repetitive activities, work responsibilities are changing to concentrate more on higher-order cognitive and creative tasks. Tasks requiring critical thinking, emotional intelligence, and problem-solving skills are becoming increasingly necessary for workers.

For instance, in the retail sector, a cashier's responsibilities are expanding beyond transaction processing to include managing client experiences and offering individualized customer care. Workers in the industrial industry are moving from operating manual assembly lines to supervising and maintaining AI-powered equipment.

The evolution of employment responsibilities often heralds the emergence of new job categories. The importance of positions like machine learning engineers, data scientists, and AI ethicists is rising. These positions call for specific expertise in AI and related fields.

Changing Requirements for Skills

AI-driven workforce transformation calls for a change in the skills that are needed. Workers need to acquire soft and technical skills to adapt to changing jobs and operate efficiently with AI technology. Soft skills are connected to AI and data analysis. Table **4** highlights the growing demand for technical skills like data analysis and AI literacy alongside soft skills such as creativity and critical thinking, which are essential for effective collaboration with AI systems.

Table 4. Changing skill requirements in an AI-driven workforce.

Importance Level	Percentage	Responses
Very Important	34.5%	~20
Important	51.7%	~30
Neutral	~3%	~2
Not Important	<1%	~1

Demanded Technical Skills

The growing need for experts who can create, implement, and oversee AI systems is reflected in the need for data science, machine learning, programming, and AI ethics. Beyond technical proficiency, there is a growing demand for AI literacy, which enables people in a variety of roles—not just in IT—to comprehend and engage with AI tools and outputs in a meaningful way.

Crucial Soft Skills

Soft talents like creativity, emotional intelligence, critical thinking, and problem-solving are equally important. These abilities allow employees to

- Apply human judgment when interpreting insights produced by AI.
- Make strategic choices in situations that are unclear and difficult.

Even while AI can process data and produce results quickly, humans are still indispensable for contextualizing findings, fostering relationships, and spurring creativity.

The results in Table **4** demonstrate the necessity of having both technical expertise to work with intelligent systems and human-centric skills to lead, plan, and innovate in the AI era. These changes necessitate updated frameworks for professional development, education, and training that give employees a well-rounded skill set for the workplace of the future.

Organizational Frameworks

AI use in the workplace is having an impact on organizational structures as well. Businesses are using more and more adaptable and agile organizational structures to react swiftly to shifts in the market and technology. Modern collaborative and decentralized models are replacing traditional hierarchical ones. Teams are now structured around specific projects or objectives, which helps them to develop and continually use AI technology efficiently. This transition calls for an environment where workers are encouraged to try new things, learn from mistakes, and be open to change.

Additionally, businesses are seeing the value of cross-functional teams that combine a range of expertise and viewpoints. For instance, data scientists, software engineers, domain specialists, and ethicists may be in a team working on an AI project. This collaborative approach ensures that the design and deployment of AI systems align with business objectives and ethical considerations.

Social and Ethical Aspects to Consider

When AI technology proliferates in the workplace, social and ethical issues must be considered. Addressing these challenges is imperative to guarantee that AI is created and applied to advance justice, fairness, and the well-being of all people.

Stopping Social and Economic Divides

Deepening social and economic divides is one of the main ethical issues with AI in the workplace. If technology is not properly regulated, the adoption of AI can worsen inequality, favoring certain groups at the expense of others. AI technology and its benefits must be accessible to everybody. This entails ensuring that every worker has access to the knowledge and training required to prosper in an AI-driven economy, irrespective of their background or socioeconomic standing. Governments, universities, and other organizations need to work together to offer accessible, reasonably priced learning options that help workers develop the skills they need. Policies and programs that support inclusion and diversity in AI-related disciplines are also crucial. Encouraging women, minorities, and low-income persons to seek professions in Artificial Intelligence (AI) can contribute to a more equitable distribution of the technology's advantages.

Ensuring the Development and Use of Ethical AI

Establishing confidence and guaranteeing the well-being of every person depends on the moral advancement and use of AI technologies. This entails creating policies and procedures that control AI systems' development, application, and use. Table **5** emphasizes the importance of ethical AI practices, including transparency in decision-making and accountability measures, to ensure fairness and build trust in AI-driven systems.

Table 5. Ethical considerations in AI deployment: Transparency and accountability.

Importance Level	Percentage	Responses
Very Important	46.6%	~27
Important	39.7%	~23
Neutral	13.6%	~8
Not Important	~0%	~0

These figures show that there is broad agreement that moral principles are necessary for the responsible and successful application of AI technologies. Given that AI is increasingly influencing important decisions across a range of industries, the overwhelming majority of respondents agree that ethical deployment is essential to maintaining justice, trust, and social responsibility.

- Key Ethical Foundations
- Openness
- AI must be comprehensible and explicable.
- This implies:
- Clearly outlining the operation of AI models, AI Transparency

To be transparent, AI systems must be comprehensible and explicable to users.

Responsibility

Making sure AI is used responsibly entails:

- Establishing decision-makers', developers', and implementers' roles and duties.
- Establishing mechanisms for ongoing audits and monitoring.
- Putting in place processes for compensation in the event that systems malfunction or cause damage.
- Mechanisms for accountability are required to identify and address biases, stop abuse, and guarantee the moral implications of AI for society.

The information in Table **5** supports the emerging understanding that moral principles need to change as technology does. Organizations must place a high priority on open communication and responsible execution if they want AI to be used in a sustainable and equitable manner. AI ethics is a social responsibility as well as a technical requirement.

Taking Discrimination and Bias Seriously

If AI systems are created without considering ethical considerations or are trained on biased data, they may unintentionally reinforce prejudice and discrimination. Prejudice in AI must be addressed to guarantee just and equal results. Organizations need strict testing and validation procedures to find and reduce biases in AI systems. This entails using representative and varied datasets, evaluating fairness, and applying methods to lessen bias, such as algorithmic tweaks and manual oversight. Additionally, encouraging diversity in AI development teams can help reduce prejudice. Diverse teams are more likely to consider various viewpoints and possible outcomes, which promotes inclusivity.

AI CAPABILITIES AND CORPORATE SOCIAL RESPONSIBILITY

AI technologies have the potential to be very helpful in fulfilling Corporate Social Responsibility (CSR) requirements. By employing personnel with AI capabilities, organizations may support the more significant social objective of driving technical innovation while guaranteeing ethical and responsible AI usage. Providing employees with training and development opportunities, funding community education campaigns, and encouraging ethical AI research are a few examples of AI-related CSR projects. Firms can also provide internships and apprenticeships to help workers and students from underprivileged communities acquire important AI skills and experience. Organizations can also collaborate with government agencies, non-profits, and academic institutions to assist research and projects that address ethical and societal challenges associated with artificial intelligence. Businesses that adopt a proactive approach to CSR may show their dedication to leveraging AI for the greater good and promoting a more inclusive and fair society.

In Conclusion, implementing AI in the workplace has benefits and drawbacks. AI-powered work automation and augmentation are changing organizational structures, skill requirements, and job responsibilities. To effectively manage these changes, workforce development must take a proactive stance that prioritizes diversity, cooperation, and ongoing learning. Addressing the social and ethical issues surrounding artificial intelligence is critical. Building trust and guaranteeing the well-being of all people depends on assuring fair access to AI technology, eliminating social and economic inequities, and supporting ethical AI development and usage. Lawmakers may use AI to create a more prosperous, just, and equitable future by focusing on these issues.

CHALLENGES BEFORE AI ADOPTION

At the vanguard of the technological revolution, Artificial Intelligence (AI) has the potential to restructure industries and change the nature of labor completely. The integration of AI into many industries has enormous potential for increasing productivity, encouraging innovation, and propelling economic growth.

Important obstacles associated with this shift must be addressed to optimize AI's advantages and reduce job displacement. This article explores adoption barriers related to AI and provides recommendations for removing them. People and organizations can manage the intricacies of this revolutionary period and create a resilient workforce capable of prospering in an AI-driven future by investing in education, guaranteeing equal access to AI developments, and creating an innovative culture.

The Price to Run an AI Model

Deep learning models need a lot of computing power to train and operate. That means expensive cloud hosting costs. Among the variables affecting cost are:

High up-front Costs

Putting AI technologies into practice sometimes necessitates a large initial outlay of funds. This covers the price of the necessary staff, hardware, software, and infrastructure.

Costs associated with cloud services: Many companies decide to store their AI models on cloud servers. Although this offers flexibility and scalability, continuous costs are associated with resource allocation and utilization.

Maintenance and Upkeep

Keeping AI models updated, maintained, and monitored increases operating expenses. It requires constant work to ensure security and good functioning.

Data Processing and Storage

Working with high-dimensional or real-time applications can result in significant costs associated with processing and storing huge datasets for training and inference.

Cost to build an AI Model

Several variables can significantly affect the price of creating an Artificial Intelligence (AI) model. Let's examine these critical factors.

Development Expenses

The main components of an AI model's development expenses are system design and construction. The total cost is influenced by the development time, the selected technology stack, and the AI application's complexity.

Costs Associated with Data

AI models depend on data, and data management, cleansing, and acquisition may be costly.

Infrastructure Costs

Another major outlay is for setting up and maintaining the required infrastructure. Hardware, software, and server hosting are examples of infrastructure costs.

Ongoing Maintenance

Routine upkeep, upgrades, and observation influence long-term expenses. It takes constant work to ensure good security and performance.

Insufficient abilities

A lack of capabilities is one of the main barriers to applying Artificial Intelligence (AI) in business. Understanding the complexities of artificial intelligence is essential for all sectors. Employees may also be concerned about losing their jobs due to AI adoption. To increase acceptance and build support, organizers should promote a collaborative and communicative culture, stress how AI helps humans and enhances their skills, and emphasize the advantages of AI.

Remember that effective AI integration requires solving the skills gap. Businesses need to make training investments, cultivate a positive workplace culture, and acknowledge AI's and humans' distinct advantages. Table **6** highlights the significance of training programs, online courses, and collaborative initiatives to bridge the skills gap and support workforce adaptation to AI technologies.

Table 6. Addressing the skills gap for AI integration.

Preparedness Level	Percentage	Number of Respondents
Very Well	24.1%	14
Well	29.3%	17
Neutral	39.7%	23
Poorly	6.9%	4

Analysis and Perspectives

According to the research, respondents' levels of readiness are uneven, with barely 25% rating good readiness and over 40% expressing uncertainty. This suggests that even if there has been improvement, a sizable section of the workforce still lacks the necessary abilities to take full use of AI technologies..

Techniques to Close the Skills Gap include training initiatives, online courses, and institutional partnerships, which are essential to closing this gap.

Initiatives for Corporate Training

By providing internal AI training that covers topics like machine learning, data analytics, and AI ethics, industry leaders like Google and IBM are establishing standards. These programs provide employee flexibility and foster a culture of lifelong learning.

Platforms for Online Learning

Platforms that provide flexible, reasonably priced, and industry-relevant content include Coursera, edX, Udacity, and others. With the help of these resources, people can autonomously upskill and maintain their competitiveness in a changing labor market.

Governmental and Academic Assistance

Academic institutions and governments have a significant influence on how prepared people are for AI.

By:

- Adding AI, data science, and emerging technologies to curricula
- Promoting STEM education at a young age and working together

The urgent need for focused action to close the skills gap in AI is highlighted in Table **6**. To create a workforce that can adjust to the demands of an AI-driven economy, governments, businesses, and educational institutions must work together. In addition to successful AI integration, strategic investments in education, training, and upskilling programs will guarantee inclusive and long-term workforce development.

Lack of Clarity Regarding Regulations

Navigating the regulatory landscape for Artificial Intelligence (AI) adoption in business presents several challenges, especially when dealing with unpredictability. Let's look into these challenges:

Complexity of Regulations

The legal and regulatory environment in which AI technologies are frequently used is complicated. Businesses may find it difficult to navigate these challenges, particularly when using AI across geographical boundaries.

Rapidly Changing Regulations

As legislators consider the moral, social, and financial ramifications of using AI, the regulatory landscape around this technology is constantly changing.

Uncertainty in Interpretation: Even when laws are established, it may not be clear how they relate to particular AI applications. Different firms may handle compliance differently due to ambiguity in regulations and varied interpretations.

International Differences

Companies that conduct business internationally must deal with both domestic and foreign regulatory frameworks. Bridging the gap between disparate regulatory systems and guaranteeing compliance internationally increases complexity.

Risk Aversion

Organisations may be more cautious when implementing AI technology due to the unpredictability of laws. Certain businesses may postpone or reduce their AI activities until regulatory uncertainties are cleared due to possible legal and reputational consequences.

Danger to Privacy of Data

Due to AI's potential to completely transform industries like manufacturing and marketing, businesses are swarming to it. However, data privacy is a significant concern associated with this rush to use AI. Below is an explanation of how the use of AI may jeopardize data privacy:

Sensitive Data Exposure

AI systems routinely consume large amounts of data, including sensitive data like names, addresses, and even bank account details. If this data is not well protected, it may be breached, which might result in identity theft and other problems.

Opaque AI Models

Even for their designers, many AI models are intricate and challenging to comprehend. Due to this "black box" effect, it is difficult to understand how the AI made a choice, which raises questions regarding prejudice and discrimination.

Data Sharing and Access by Third Parties

Businesses frequently exchange data with other providers to train and maintain AI models. This increases the possibility that information may be abused or end up in the wrong hands.

Data Retention and Deletion

Understanding how long data is kept and when it should be safely erased is critical. Businesses must have well-defined policies to comply with data privacy laws. Sensitive data inference: Even if you don't give sensitive data to an AI directly, it may be able to deduce it from other data. For instance, an AI that studies consumer behavior could infer someone's income or medical issues.

USERS' EDUCATION TO COMPREHEND PRODUCT LIMITATIONS AND AI USE CASES

The education of users about the limitations of AI is one of the main obstacles to its adoption by enterprises. Even though AI has much promise, users must know its limitations to guarantee responsible and successful use. Here are some issues that might arise from this lack of user education:

Unrealistic Expectations

Users who assume AI is perfect may rely too much on its results and fail to see biases or flaws. This can undermine confidence in the technology and result in bad decision-making.

Misuse of AI Tools

Users may undertake tasks for which the AI was not intended if they are unaware of the limitations of the AI's capabilities. This may lead to missed opportunities and erroneous findings.

Opposition to Change

People uneasy with AI because they don't know enough about it may be reluctant to accept it, limiting the technology's potential advantages for the company.

MORAL ISSUES IN THE ADOPTION OF AI

The incorporation of AI into numerous facets of society raises significant ethical questions. It is crucial to ensure that AI technologies are created and applied in a way that advances justice, fairness, and everyone's well-being. Fairness, prejudice reduction, and privacy are important ethical concerns.

PRIVACY ISSUES

Massive volumes of data are frequently required for AI systems to work well. This dependence on data presents serious privacy problems because personal information about individuals can be collected. One of AI's most important ethical challenges is ensuring data privacy and safeguarding individual rights. Organizations must have strong data protection mechanisms, such as encryption, anonymization, and secure data storage, to protect personal information. Establishing explicit and transparent data usage policies is also necessary to get people's consent and tell them how their data is used. The European Union's regulations, such as the General Data Protection Regulation (GDPR), provide a framework for safeguarding data privacy. Compliance with these standards is crucial to fostering confidence and guaranteeing the responsible use of AI technology.

EQUITY AND BIAS ELIMINATION

If AI systems are not carefully planned and executed, they have the potential to reinforce preexisting prejudices and make them worse. Biases in artificial intelligence might result from algorithm inaccuracies, biased training data, or unintentional system design repercussions. Promoting equity and preventing discriminatory outcomes requires ensuring fairness and minimizing bias in AI systems.

Organizations must take a multipronged approach to bias management, which involves collecting representative and varied data, rigorously testing and validating AI systems, and continuously checking for bias. Strategies like algorithmic fairness interventions and human-in-the-loop methods can find and lessen bias.

It's also critical to encourage diversity in AI development teams. Diverse teams may uncover possible biases and create more egalitarian AI systems by bringing a variety of viewpoints and experiences to the table. Programs to promote underrepresented groups in AI-related occupations and broaden diversity in STEM sectors can facilitate this.

ACCOUNTABILITY AND TRANSPARENCY

Accountability and transparency are fundamental to moral AI. Transparency in designing and implementing AI systems is essential so users may question results and comprehend the decision-making process. It is imperative to establish accountability systems that hold AI developers and users accountable for the outcomes of an AI deployment.

Organizations may encourage openness by concisely explaining AI algorithms, decision-making procedures, and data usage. Furthermore, accountability frameworks and audit trails can ensure that AI systems are used responsibly and that any negative effects are quickly remedied.

AI SUGGESTIONS IN AI ADOPTION

Organizations and people must manage several hurdles while using AI technology. Proactively tackling these issues makes it feasible to optimise AI's advantages and create a resilient workforce that can prosper in an AI-driven future.

Workplace Shift and Employment Losses

One of the biggest obstacles is the possible loss of jobs due to AI adoption. As regular tasks are automated by AI systems, some occupations may become obsolete, raising worries about unemployment and economic upheaval. To solve this dilemma, it is crucial to concentrate on workforce transformation and the establishment of new employment possibilities.

Programs for Upskilling and Reskilling

Programs for upskilling and reskilling employees are essential for assisting them in adjusting to new responsibilities in an AI-driven economy. Workers should graduate from these programs with the knowledge and abilities required for new career paths like digital literacy, data analysis, and AI creation.

Table 7 demonstrates that efforts to reskill and upskill employees are not only highly regarded but also necessary to create a workforce prepared for the future. Coordination between governments, corporations, and educational institutions is essential to maximizing their influence. To guarantee that employees are prepared to handle the opportunities and difficulties brought about by AI-driven changes, these programs need to be innovative, inclusive, and dynamic.

Table 7. Importance of reskilling and upskilling in preparing the workforce for AI-driven transformations.

Usefulness Level	Percentage	Number of Respondents
Very Useful	46.6%	27
Useful	32.8%	19
Neutral	15.5%	9
Not Useful	5.2%	3

Assistance for Relocated Employees

It is important to offer assistance to displaced workers to facilitate a seamless transition to alternative employment prospects. Financial aid for school and training, career counseling, and job placement services are examples of this help.

Organizations may provide all-encompassing assistance to people in navigating the difficulties of job displacement and discovering new possibilities in the AI-driven economy.

Encouraging Creativity and Efficiency

AI can significantly increase productivity and innovation in various sectors. However, to fully realize this promise, a number of obstacles must be overcome, such as a lack of infrastructure, reluctance to adapt, and restricted access to AI technology.

Developing a Creative Culture

Organizations must foster an innovative culture to utilize Artificial Intelligence (AI) fully. This entails establishing a setting that values creativity, promotes experimentation, and views failure as a teaching tool. Organizations may harness AI technology by cultivating a creative culture to encourage innovation and inspire staff to explore new ideas.

Leaders have a critical responsibility to foster an innovative culture. They may set the example by promoting invention, offering tools and encouragement for trial, and praising and rewarding creative work. Moreover, assembling cross-functional teams with various backgrounds and viewpoints may foster innovation and lead to ground-breaking ideas.

Putting Money into Infrastructure

For AI adoption to be successful, the infrastructure must be invested in. This covers cloud computing services, fast internet, and sophisticated data processing and storage tools. The infrastructure required to enable AI technologies must be built and maintained, and this requires funding from governments and other organizations. Encouraging fair access to AI innovations can also be significantly aided by public infrastructure investment. Governments can help close the digital divide and guarantee that everyone can use AI technology by ensuring that underprivileged and rural populations have access to high-speed internet and other necessary resources.

Providing Fair Access to AI Developments

Fair access to AI developments is essential to halting the widening of social and economic divides. Promoting equality and inclusion requires ensuring that everyone has access to AI technology and the possibilities it offers, regardless of their background or socioeconomic situation.

Accessible and Economical Education

Ensuring fair access to AI developments requires affordable and accessible education. Governments, academic institutions, and organizations must collaborate to offer reasonably priced educational options that allow people to gain the necessary competencies for AI-related careers.

Encouraging Inclusion and Diversity in AI

Encouraging inclusivity and diversity in AI-related domains is crucial to guaranteeing that AI's advantages are broadly shared. Supporting underrepresented groups, including women, minorities, and those from low-income backgrounds, may help achieve a more inclusive AI ecosystem. Organizations can implement diversity and inclusion initiatives that help underrepresented groups advance their careers by offering networking opportunities, career development support, and mentoring. Furthermore, encouraging diversity in AI development teams helps guarantee that various viewpoints and experiences are incorporated into the design of AI systems, producing more equitable results.

In conclusion, the Adoption of AI technology comes with obstacles as well as possibilities. Organizations and individuals may use AI to boost productivity and spur creativity by proactively tackling these issues.

FRAMEWORKS AND GLOBAL BENCHMARKS FOR AI-DRIVEN WORKFORCE TRANSFORMATION

Table 8. Summary of major global reports on skills development and digital education policies.

Framework	Important Attention Areas	Brief Explanation	Citation
OECD Skills Strategy (2019, 2021)	Digital skills, inclusivity, and lifelong learning	Suggests enhancing education quality, creating flexible learning systems, and encouraging fair access to chances for upskilling.	OECD (2019, 2021) https://www.oecd.org/skills
WEF Future of Jobs Report (2023)	Partnerships, future skills, and reskilling	Forecasts a 44% disruption in skills by 2027 and calls for funding for public-private reskilling initiatives, emotional intelligence, and creativity.	WEF (2023) https://www.weforum.org/reports/the-future--f-jobs-report-2023
UNESCO Futures of Education (2021)	ethical AI Collaboration and digital competencies	Calls for inclusion, AI ethics, and digital literacy to be incorporated into future-ready education.	UNESCO (2021) https://unesdoc.unesco.org/ark:/48223/pf0000379707
EU Digital Education Action Plan (2021–2027)	Reimagined education, equity	Intends to raise the degree of digital proficiency and foster close collaboration between academic institutions, businesses, and schools.	European Commission (2020) https://education.ec.europa.eu
UNDP Human Development Report (2021/2022)	Equity in technology and inclusive growth	Emphasizes balanced investment in digital tools and education, as well as human-centered technology development.	UNDP (2022) https://hdr.undp.org

CONCLUSION

The emergence of Artificial Intelligence (AI) represents a turning point in the development of the world's labor force. The need for proactive adaptation and ongoing learning is becoming increasingly apparent as AI continues to enter all economic areas (Table **8**). This conclusion emphasizes the value of ethical issues, the need for upskilling and reskilling, and a cooperative approach to human-AI integration. By adhering to these guidelines, people and institutions may turn obstacles into chances and develop a resilient labor force that can prosper in an AI-driven society.

Table 9. Strategic collaborations for developing AI-ready workforce ecosystems.

Support Level	Percentage	Number of Respondents
Very Supportive	20.7%	12
Supportive	60.3%	35
Neutral	17.2%	10
Not Supportive	1.7%	1

The claim that cooperation between humans and AI is viewed as a beneficial and essential path for the future development of the labor market is supported by Table **9**. Human-AI integration has enormous potential to build a resilient, future-ready, and just global workforce when paired with solid ethical underpinnings, a strong training infrastructure, and inclusive support networks. As technology advances quickly, upskilling and reskilling become essential tactics for people and businesses looking to maximize AI's potential while lowering the possibility of job displacement. Because AI technologies are dynamic, we need a workforce that is not just technically skilled but also flexible and eager to learn new things throughout their lives. Workers who place a high priority on lifelong learning and skill development may easily move into new professions that use AI to boost creativity and productivity.

Establishing a culture of continual learning and funding staff training programs is crucial for organizations. This investment improves organizational agility and competitiveness and prepares the staff for future needs. Government programs, public-private alliances, and scholarly partnerships may further support these efforts, ensuring that the workforce has the skills required to navigate an AI-driven economy. Justice, Privacy, and Human-AI Collaboration as Ethical Considerations. AI's introduction into the workplace raises critical ethical issues that must be resolved to guarantee that technology is used responsibly and for the greater good. Concerns of equity, confidentiality, and minimization of prejudice are critical to creating and applying AI systems. It's essential to ensure that AI technologies are developed and applied in a way that upholds justice and equity to avoid making already-existing social and economic inequalities worse.

Privacy concerns are significant, with AI systems requiring large amounts of data. Companies need to implement strong data security and protection protocols, and clear data usage guidelines are essential for organizations to preserve people's privacy and foster confidence. To detect and reduce biases and make sure that AI systems do not support discrimination or unjust behaviors, ethical AI development also includes thorough testing and validation.

Another essential component of integrating AI ethically is human-AI cooperation. It's important to see AI's potential to enhance human skills rather than see it as a replacement for human labour. Organisations can improve innovation, creativity, and decision-making by promoting a collaborative atmosphere where people and AI work together. In addition to optimizing AI's advantages, this cooperative strategy ensures that workers continue to be essential to the process of creating value.

Lifelong learning, diversity, and creativity are the cornerstones of a resilient workforce.

A dedication to diversity, innovation, and continual learning will be essential as we prepare for an AI-driven future and turn obstacles into possibilities. Diversity in the workforce and development teams for AI guarantees that a wide range of viewpoints and experiences are taken into account, which results in AI systems that are more inclusive and egalitarian. Supporting underrepresented groups in STEM and AI-related professions can contribute to developing a more robust and diverse workforce.

Creativity and innovation are equally significant when it comes to adopting AI. Organizations may promote continuous development and remain ahead of technology breakthroughs by encouraging people to explore new ideas and cultivating a culture of experimentation. Leaders are essential in fostering this culture because they give resources, encouragement, and credit for creative work. A resilient workforce is built on a foundation of lifelong learning. As AI technologies advance, new competencies will always be needed. Workers must be encouraged to pursue ongoing education and skill development to stay relevant in their jobs. Flexible learning alternatives like micro-credentials and online courses may help people better manage their education with their employment and other commitments.

Creating Opportunities Out of Obstacles

Adopting AI is not tricky but offers special chances for advancement and progress. Organizations may create chances for innovation and competitiveness by taking a proactive approach to workforce development. Important milestones in this process include funding training and instruction, advocating for moral AI practices, and cultivating an inclusive and cooperative culture.

Ensuring that AI's advantages are widely spread requires equitable access to breakthroughs in the field. Governments, universities, and other organisations must collaborate to offer accessible and reasonably priced educational opportunities so that everyone may acquire the knowledge and abilities required for careers in artificial intelligence. By working together, we can ensure that underprivileged communities are not left behind and help close the digital gap.

Another critical component of transforming challenges into opportunities is fostering diversity and inclusivity in AI-related disciplines. By assisting underrepresented groups and providing them with paths into AI, organizations may gain access to a larger pool of talent and opportunities for career advancement.

Establishing a Resilient Staff

The ultimate objective is to create a robust labor force that can successfully negotiate the challenges of an AI-driven society. This necessitates a comprehensive strategy that includes ongoing education, moral reflection, and a dedication to tolerance and diversity. By adopting these principles and preparing for the future of work, individuals and organizations can ensure that AI is a force for good change in the workplace.

As we advance, it is critical to understand that the future of work is shaped by our decisions about how to best utilize technology, not only by it. By welcoming diversity and inclusion, emphasizing ethical issues, and making investments in workforce development, we can guarantee that AI acts as a catalyst for a fairer, egalitarian, and affluent society. To sum up, the skills revolution brought about by AI necessitates proactive adaptation, ongoing learning, and a dedication to moral behavior. We can create a robust and successful workforce in an AI-driven future by tackling the difficulties and taking advantage of the possibilities that AI presents. The road ahead is long and winding, but if we work together, we can make the most of the challenges and build a future where AI powers creativity and human cooperation.

REFERENCES

ARK Investment Management LLC. (n.d.). Stanford DAWN Deep Learning Benchmark (DAWNBench). Available from: https://arkinv.st/2MUITyO.

Aquino, Y.S.J., Rogers, W.A., Braunack-Mayer, A., Frazer, H., Win, K.T., Houssami, N., Degeling, C., Semsarian, C., Carter, S.M. (2023). Utopia versus dystopia: Professional perspectives on the impact of healthcare artificial intelligence on clinical roles and skills. *Int. J. Med. Inform., 169*, 104903. [http://dx.doi.org/10.1016/j.ijmedinf.2022.104903] [PMID: 36343512]

Butler, D. (2016). Tomorrow's world: Technological change is accelerating at an unprecedented speed and could create a world we can barely imagine. *Nature, 530*(7591), 398-402. [http://dx.doi.org/10.1038/530398a] [PMID: 26911763]

Bower, M. (2019). Technology-mediated learning theory. *Br. J. Educ. Technol., 50*(3), 1035-1048. [http://dx.doi.org/10.1111/bjet.12771]

Coombs, C., Hislop, D., Taneva, S.K., Barnard, S. (2020). The strategic impacts of Intelligent Automation for knowledge and service work: An interdisciplinary review. *J. Strateg. Inf. Syst., 29*(4), 101600. [http://dx.doi.org/10.1016/j.jsis.2020.101600]

Chui, M., Manyika, J., Miremadi, M. Four fundamentals of workplace automation. *McKinsey Q., 29*(3), 1-9.Available from: https://www.mckinsey.com/capabilities/mckinsey-digital/our-insights/four-fundament-ls-of-workplace-automation. (2015).

Jaiswal, A., Arun, C.J., Varma, A. (2022). Rebooting employees: upskilling for artificial intelligence in multinational corporations. *Int. J. Hum. Resour. Manage., 33*(6), 1179-1208. [http://dx.doi.org/10.1080/09585192.2021.1891114]

Jovanovic, M., Campbell, M. (2022). Generative artificial intelligence: Trends and prospects. *Computer, 55*(10), 107-112. [http://dx.doi.org/10.1109/MC.2022.3192720]

Kunnen, E.S., Bosma, H.A. (2003). Fischer's skill theory applied to identity development: A response to Kroger. *Identity, 3*(3), 247-270. [http://dx.doi.org/10.1207/S1532706XID0303_05]

CHAPTER 8

Examining the Role of IT Industries Towards Work Compatibility and the Well-being of Women Employees in Chennai City

T. Shalini Priyadharshini[1] and **D. Anitha Kumari**[1,*]

[1] *School of Management Studies and Commerce, Vels Institute of Science, Technology and Advanced Studies (VISTAS), Pallavaram, Chennai-117, Tamil Nadu, India*

Abstract: The constant expansion of the information technology industry and globalization have substantially changed the corporate landscape. Not only has it altered the dynamic between employers and employees, but it has also changed how business is conducted and what is expected of employees. The management of a firm has the primary responsibility for its performance and profitability, owing to today's diversified workforce and demanding environment. Maintaining a healthy work-life balance in today's challenging and competitive workplace is essential. Employees' personal and professional lives need to be balanced for work-life balance. A sample of 150 female employees from the IT industry was utilized to investigate how women perceive work compatibility and how it influences their well-being.

The inquiry employed a quantitative methodology. A structured questionnaire was distributed to 200 respondents to gather data. One hundred fifty only returned the fully completed questionnaire. After that, this data was examined using the appropriate statistical programs. According to the quantitative approach, the data is found to be quantitative and falls into three categories: simulation, experimentation, and inferential. The trial form contains a characteristic that provides a great deal of adaptability in the research setting: the ability to modify some of the variables to observe how they impact the other variables. Creating a synthetic setting in which valuable data and details may be made is one of the most crucial components of the simulation process. The study's findings demonstrate that healthy work compatibility boosts employee well-being, which increases organizational effectiveness and gives IT companies a competitive edge.

Keywords: Motivation, SEM model, Work compatibility, Women employees, Well-being.

* **Corresponding author Anitha Kumari:** School of Management Studies and Commerce, Vels Institute of Science, Technology and Advanced Studies (VISTAS), Pallavaram, Chennai-117, Tamil Nadu, India; E-mail: anitha.sms@velsuniv.ac.in

Rishikaysh Kaakandikar, Keshav Kaushik, Supriya Pathakm Madhavi Shamkuwar & Vivek Mishra (Eds.)

INTRODUCTION

The workplace has never been the same, thanks to the evolution of technology. Today's organizations are developing technically and in terms of personnel, experience, and age. The information technology industry is no different. Millennials are becoming a noticeable presence in the information technology industry. They are very ambitious and focused on their careers. Due to its numerous options, competitive pay, and global exposure, this industry draws many millennials. The traditional view of men as the breadwinners and women as homemakers or child bearers has changed to view women as capable of undertaking domestic duties and being the breadwinners. More women than ever before are employed outside these days. Particularly, there are now more millennial women employed in the information technology industry; however, this is not always the case because the job in this field is challenging and never seems to stop. Most IT workers are constantly under extreme pressure to fulfil customer demands, supervisor expectations, and the elevated internal requirements they establish for themselves. The number of women working in this field has raised the importance of work-life balance. Not only have millennial women workers embraced a gender-neutral workplace, but they have also become more aware of the difficulties posed by competing expectations and different job demands. Millennial women have entered this field, but they still have other family duties, such as nurturing. The shifting expectations and aspirations of millennial women employees, such as marriage and children at different stages of their lives, influence work-life compatibility. Thus, attaining work-life harmony in this unstable work environment has become risky. This study aims to ascertain the issues millennial women employees have regarding work-life balance.

Research on work-life balance and gendered roles across various industries has been a focus in recent years. Chauhan *et al.* (2022) explore how family responsibilities, mentoring, and organizational support contribute to career success for women. Similarly, Arora and Vally (2018) examine the challenges workers face in IT companies in India, particularly in balancing work and personal life. Gendered ideologies surrounding work-life balance in the IT sector have been studied by Heikkinen and Kivijärvi (2022), who found that gender plays a significant role in shaping work-life expectations. Additionally, Geetha and Ashtikar (2022) highlight the importance of work-life balance among women in private higher education institutions during the COVID-19 pandemic and its relation to organizational sustainability. The impact of Generation Y employees on IT organizations has also been discussed by Krishna Kumar and Velmurugan (2018), emphasizing how younger employees influence organizational dynamics. Gendered perceptions of work in the IT sector have been compared by Pyrkosz-Pacyna, Dukala, and Kosakowska-Berezecka (2024), while Croft *et al.* (2015)

delve into the cultural and psychological barriers that hinder men's engagement in communal roles. The broader implications of work-life balance and its effect on career and personal choices are also explored in academic settings, as shown by Aelenei *et al.* (2020), who discuss how an educational culture based on self-enhancement values can impact female students' sense of belonging and academic aspirations. These studies collectively contribute to understanding the complex interplay between work-life balance, gender, and organizational success.

Research Problem

The field of information technology is expanding. The millennial generation, crucial to this industry, is tech-aware and has distinct expectations and beliefs from earlier generations in the workplace. A formidable obstacle to attaining work-life balance for women of this age was the introduction of millennial women into this field. Demands at work also rise for millennial women as they reach new life stages, including marriage, parenthood, and numerous other life commitments. As a result, people are obliged to choose between their personal and professional objectives. Millennial women with strong job goals and great aspirations will struggle to balance work and personal obligations. Therefore, the purpose of this study was to ascertain the worries of Women of the millennial age employed in the data technology industry.

Need for the Study

With women making up almost half of the workforce, the Millennial Generation is currently the largest generational cohort in the Indian workforce. Enterprises may create the kinds of workplaces that many employees want by knowing the policies and procedures that best empower various groups. This will improve their capacity to attract and retain brilliant workers, which will eventually make them more successful enterprises. But there are a lot of misconceptions regarding Millennials and women in the workforce. These groups frequently work for organizations that do not recognize or value their skills and requirements. Many businesses have tried to develop programs that empower women, Millennials, or both as these generations have joined the workforce at accelerating rates. Businesses must have access to evidence-based information on the unique requirements, difficulties, and preferences of these groups if these programs are to be successful. Additionally, by identifying areas of overlap, this kind of study may give businesses insightful advice on how to empower both Millennials and women.

Objectives of the Study

- To know the demographic profile of the women employees working in the selected IT companies, Chennai city.
- To explore the factors influencing the work compatibility of women employees working in the selected IT companies, Chennai city.
- To validate the relationship between the working compatibility of women IT employees in Chennai city and their well-being.

Theoretical Background and Hypotheses Development

Chauhan, J., Mishra, G., and Bhakri, S. (2022) stated that although women's engagement and empowerment are growing and gender inequality is diminishing, women are still represented in high-level jobs in India. Women face several obstacles in pursuing senior or top-level roles in the workforce. Since a woman must overcome several hurdles to achieve job success, this research focuses on the challenges women face due to organizational and familial restraints. The current study aims to ascertain if women's perceptions of their professional success in the Indian information technology sector are influenced by perceived organizational support, family obligations, and a lack of mentorship. A systematic questionnaire was utilized to gather information in this respect, and 292 respondents were examined. The structural equation modelling method investigated how these obstacles affected women's views of their job success. The study's findings show that every independent variable significantly impacts how successful women see their employment, which should be a wake-up call for female executives who need to overcome these barriers to advance in their careers.

Krishna Kumar *et al* (2018) surveyed how well millennial workers in Cochin's information technology industry managed their work-life balance. It identified the factors influencing the work-life balance of millennials. Studies show that workers under 25 are not as good at balancing work and personal commitments as millennials between the ages of 25 and 35. This study also discovered that workers skilled at managing their work-life balance include those who are married, have rotational schedules, are trainees, have graduate degrees in IT, and are knowledgeable about work-life balance. Moreover, these workers' socioeconomic backgrounds significantly influence how healthy the work-life balance is.

Jyonna Pyrkosz-Pacyna (2024) found that community objectives are more critical to IT employees than to IT students and that female IT professionals felt more socially connected than both male and female IT students. Women in non-IT groups reported a similar sense of social belonging to that of women working in IT. These results imply that women's perceptions of the IT field may change and become more in line with their demands once they start working there. Additionally, we discovered that whereas female IT workers valued social objectives more than agentic goals, female students did not share this sentiment. The findings emphasize how crucial it is to examine how women view the IT industry regarding their ambitions and other work-related factors at various career stages. These research avenues will contribute to developing more potent interventions to encourage women to choose careers in IT.

Hypotheses Development

- $H_0$1: There is no significant impact of skill and qualifications on well-being
- $H_0$2: There is no significant impact of interests and motivations on well-being
- $H_0$3: There is no significant impact of Personality and work style on well-being
- $H_0$4: There is no significant impact of Values and culture fit on well-being

METHODOLOGY

This study, which is quantitative in nature, makes use of both primary and secondary data. A questionnaire is used to obtain primary data for this research. Respondent answers are gathered using a convenience sample technique. 150 millennial women working in the field of information technology in Chennai made up the study's respondents. This study gathers secondary data from books, journals, websites, and other sources. With the use of SPSS, the basic data that were collected were examined. Exploratory factor Analysis and Structural Equation Modelling were performed to determine millennial women workers' worries about work compatibility and well-being.

Analysis and Results

Table **1** presents the demographic profile of respondents, including age, educational qualifications, marital status, number of children, type of family, and family size. Most respondents are 41–50 years old, are UG graduates, and are married. It highlights family structure and size diversity, with nuclear families slightly more prevalent than joint families. This demographic distribution provides a comprehensive overview of the study's participant characteristics.

Table 1. Demographic profile of respondents.

Age	No. of. Respondents	Total Percentage
below 30	23	15.3
31 – 40 years	37	24.6
41-50 years	55	36.7
Above 50 years	35	23.4

(Table 1) cont.....

Age	No. of. Respondents	Total Percentage
Total	**150**	**100.0**
Educational Qualification	**No. of. Respondents**	**Total Percentage**
UG	75	50
PG	32	21.4
Ph.D.	28	18.6
Others	15	10
Total	**150**	**100.0**
Marital Status	**No. of. Respondents**	**Total Percentage**
Single	55	36.6
Married	62	41.4
Others	33	22
Total	**150**	**100.0**
Number of Children	**No. of. Respondents**	**Total Percentage**
None	58	38.6
1	35	23.4
2	49	32.6
More than 2	8	5.4
Total	**150**	**100.0**
Type of family	**No. of. Respondents**	**Total Percentage**
Nuclear	82	54.6
Joint	68	45.4
Total	**150**	**100.0**
Family size	**No. of. Respondents**	**Total Percentage**
Less than 2 members	38	25.4
2-4 members	60	40
4-6 members	34	22.6
Above 6 members	18	12
Total	**150**	**100.0**

It is evident from Table **1** that 36.7% of respondents are in the age range of 41 to 50 years, 24.6% are in the age group of 31 to 40 years, and 23.4% and 15.3% are in the age categories of above 50 years and below 30 years, respectively. It is clear that the bulk of respondents 50% are UG graduates, followed by PG graduates (21.4%), Ph.D. holders (18.6%), and other respondents (10%). Regarding marital status, 41.4% of respondents, 41.4% are married, followed by single respondents (36.6%) and others (22%). Regarding the heads of number of children, 38.6% of respondents have no children, 32.6% have two, 23.4% have one, and 5.4% have more than two. Regarding the heads of family type, most respondents (54.6%) are in nuclear families, while 45.4% are in joint families. According to the family size heads, 40% of families consist of two to four members, 25% have fewer than two members, 22.6% have four to six members, and 12% have more than six individuals.

Table **2** shows that all the mean values are above three according to the guideline, and all the standard deviation values are above 7.

Table 2. Item statistics for skill and qualifications.

-	Mean	Std. Deviation	Cronbach's Alpha if Item Deleted
Whether the position is entry-level, intermediate, or advanced, I believe that the level of proficiency needed for it	3.84	.994	.913
Working with my company requires effective communication, emotional intelligence, and flexibility.	3.72	1.057	.935
Experience and technical proficiency may take precedence over formal education in some positions.	3.79	1.034	.894

Table **3** shows that all the mean values are above three according to the guideline, and all the standard deviation values are above 7.

Table 3. Item Statistics for personality and work style.

-	Mean	Std. Deviation	Cronbach's Alpha if Item Deleted
Assessments can give important details about a person's problem-solving methodology, communication style, and attitude toward teamwork.	3.68	1.113	.922
Evaluating cultural fit can assist in averting future disputes.	3.82	.963	.776
Make sure our working style is appropriate for the job.	3.70	.988	.792
Proficiency in dispute resolution is essential, particularly in positions where collaboration is essential.	3.82	.943	.776

Table **4** shows that all the mean values are above three according to the guideline, and all the standard deviation values are above 7.

Table 4. Item Statistics for values and Culture Fit.

-	Mean	Std. Deviation	Cronbach's Alpha if Item Deleted
Inquire more to learn about their priorities, ethical standards, and true concerns in the workplace.	3.83	1.053	.810
I feel free to share my experiences working in comparable settings.	3.78	.944	.835
I'm outlining my goals, principles, and objectives while also giving examples of how I live them out every day.	3.71	1.005	.831

Table **5** shows that, according to the guideline, all the mean values are above three, and all the standard deviation values are above 7.

Table 5. Item statistics for interests and motivations.

-	Mean	Std. Deviation	Cronbach's Alpha if Item Deleted
I provide a task or project that fits with my interests in my line of work.	3.19	1.162	.841
Employers sync my motivations with the objectives of the company and the job.	2.98	1.193	.848
Company is able to offer the required chances for growth.	3.34	.993	.874
My employer has career development plans, recognition programs, and continuous feedback systems in place.	3.36	1.050	.853

Table 6. Item statistics for well-being.

-	Mean	Std. Deviation	Cronbach's Alpha if Item Deleted
Training Programs	4.15	.933	.865
Mentorship programs	3.89	1.008	.886
Feedback Culture	4.04	1.055	.845
Cross-Training	3.97	1.076	.860

Table **6** shows that, according to the guideline, all the mean values are above three, and all the standard deviation values are above 7.

Table **7** reveals that the five constructs, comprising 18 items, cumulatively explain 79.454 per cent of the total variance.

Table 7. Total variance explained.

Component	Initial Eigenvalues			Extraction Sums of Squared Loadings			Rotation Sums of Squared Loadings		
	Total	% of Variance	Cumulative %	Total	% of Variance	Cumulative %	Total	% of Variance	Cumulative %
1	3.759	20.882	20.882	3.759	20.882	20.882	3.208	17.824	17.824
2	3.086	17.147	38.029	3.086	17.147	38.029	3.057	16.983	34.807
3	3.045	16.919	54.948	3.045	16.919	54.948	3.020	16.777	51.584
4	2.365	13.136	68.084	2.365	13.136	68.084	2.600	14.446	66.030
5	2.046	11.369	79.454	2.046	11.369	79.454	2.416	13.424	79.454
6	.671	3.727	83.181	-	-	-	-	-	-
7	.585	3.249	86.430	-	-	-	-	-	-
8	.498	2.769	89.199	-	-	-	-	-	-
9	.321	1.782	90.980	-	-	-	-	-	-
10	.297	1.651	92.631	-	-	-	-	-	-
11	.254	1.411	94.042	-	-	-	-	-	-
12	.252	1.398	95.440	-	-	-	-	-	-
13	.181	1.006	96.446	-	-	-	-	-	-
14	.161	.894	97.340	-	-	-	-	-	-
15	.145	.804	98.145	-	-	-	-	-	-
16	.126	.698	98.843	-	-	-	-	-	-
17	.115	.639	99.482	-	-	-	-	-	-
18	.093	.518	100.000	-	-	-	-	-	-
Extraction Method: Principal Component Analysis.									

The component loadings of various work compatibility and well-being items were analyzed through factor analysis. As shown in Table **8**, five components were identified, each representing distinct constructs: Skill and Qualifications (Component 1), Well-Being (Component 2), Interests and Motivations (Component 3), Personality and Work Style (Component 4), and Values and Culture Fit (Component 5). Notably, 'Experience and technical proficiency may take precedence over formal education' and 'Feedback Culture' exhibited high loadings of 0.955 and 0.901, respectively, underscoring their strong representation within their respective components. These results highlight the multifaceted factors influencing work compatibility among millennial women in the IT sector.

Table 8. Rotated component matrix[a].

-	Component				
	1	2	3	4	5
Experience and technical proficiency may take precedence over formal education in some positions.	.955	-	-	-	-
Whether the position is entry-level, intermediate, or advanced, I believe that the level of proficiency needed for it	.935	-	-	-	-
Working with my company requires effective communication, emotional intelligence, and flexibility.	.883	-	-	-	-
Assessments can give important details about a person's problem-solving methodology, communication style, and attitude toward teamwork.	.758	-	-	-	-
Feedback Culture	-	.901	-	-	-
Cross-Training	-	.877	-	-	-
Training Programs	-	.875	-	-	-
Mentorship programs	-	.835	-	-	-
I provide a task or project that fits with my interests in my line of work.	-	-	.881	-	-
Employers sync my motivations with the objectives of the company and the job.	-	-	.874	-	-
My employer has career development plans, recognition programs, and continuous feedback systems in place.	-	-	.871	-	-
Company is able to offer the required chances for growth.	-	-	.830	-	-
Proficiency in dispute resolution is essential, particularly in positions where collaboration is essential.	-	-	-	.938	-
Evaluating cultural fit can assist in averting future disputes.	-	-	-	.924	-
Make sure our working style is appropriate for the job.	-	-	-	.893	-
Inquire more to learn about their priorities, ethical standards, and true concerns in the workplace.	-	-	-	-	.907
I'm outlining my goals, principles, and objectives while also giving examples of how I live them out every day.	-	-	-	-	.894
I feel free to share my experiences working in comparable settings.	-	-	-	-	.885

Factor 1

Every item that loaded higher than the permissible threshold of 0.05 or 0.70 and those with low factor loading were removed from the analysis. Consequently, every one of these items constitutes a distinct construct named **Skill and Qualifications.**

Factor 2

Every item that loaded higher than the permissible threshold of 0.05 or 0.70 and those with low factor loading were removed from the analysis. Consequently, every one of these items constitutes a distinct construct named **well-being.**

Factor 3

Every item that loaded higher than the permissible threshold of 0.05 or 0.70 and those with low factor loading were removed from the analysis. Consequently, every one of these items constitutes a distinct construct named **Interests and Motivations.**

Factor 4

Every item that loaded higher than the permissible threshold of 0.05 or 0.70 and those with low factor loading were removed from the analysis. Consequently, each of these items constitutes a distinct construct called **personality and work style.**

Factor 5

Every item that loaded higher than the permissible threshold of 0.05 or 0.70 and those with low factor loading were removed from the analysis. Consequently, every one of these items constitutes a distinct construct named **values and Culture Fit.**

Confirmatory Factor Analysis

The initial model (Fig. **1**) demonstrated suboptimal fit indices, as indicated in Table **9**. The results revealed that modifications were necessary to align the model with theoretical expectations. The issues in the initial model prompted adjustments to address error correlations and achieve a better level of acceptance.

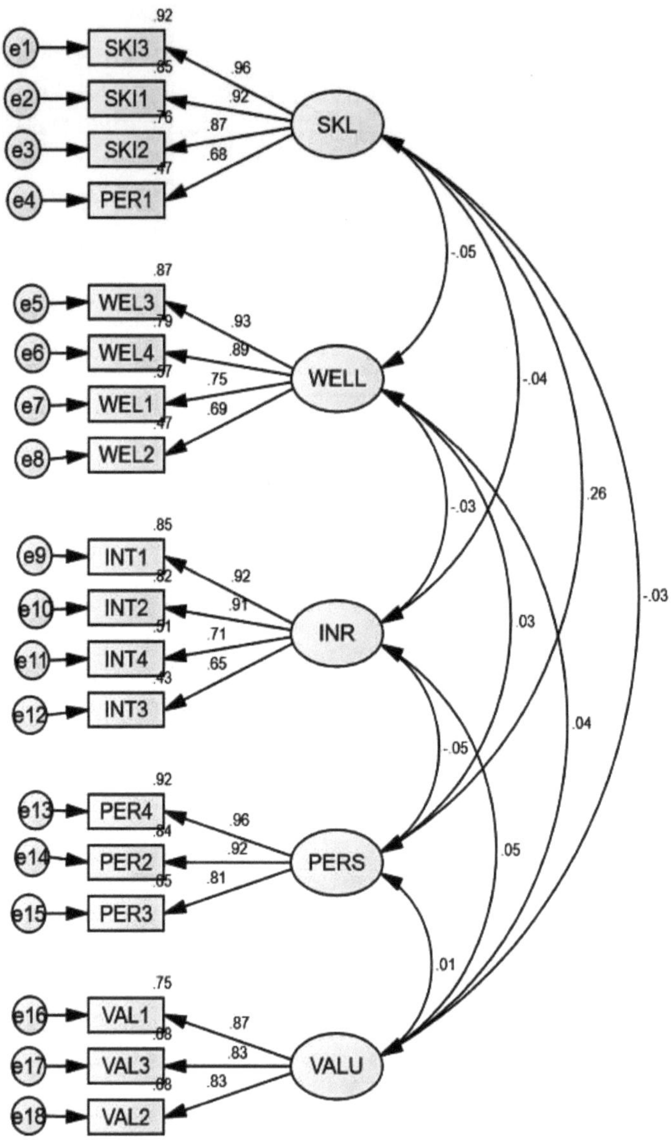

Fig. (1). Initial model.

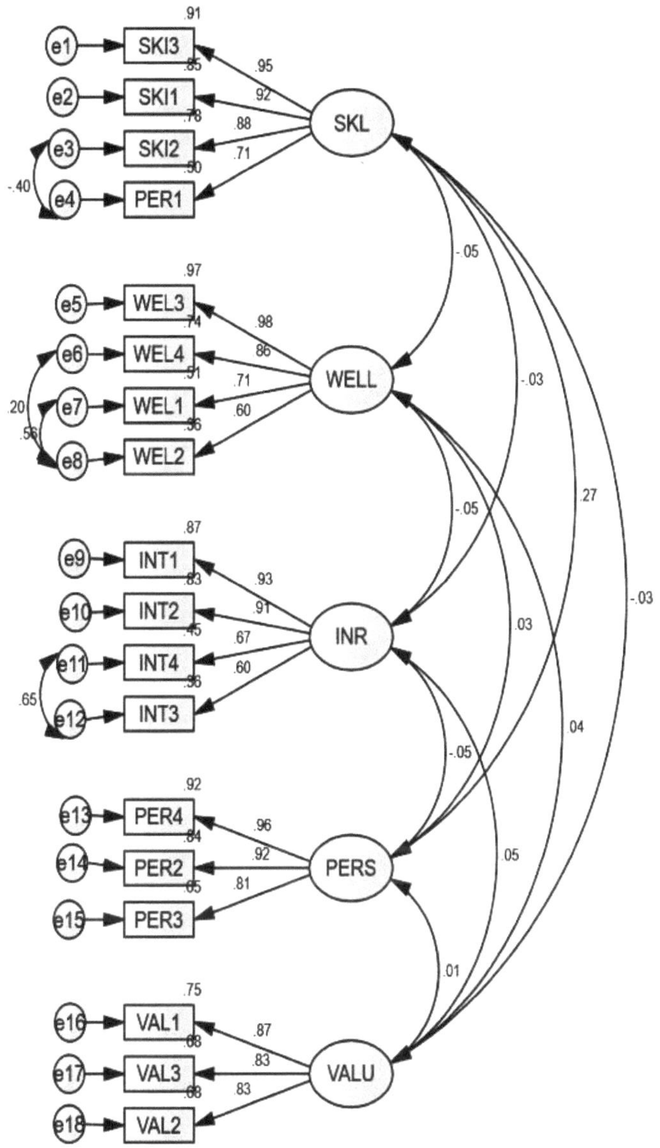

Fig. (2). Modified model.

Table 9. Fit statistic change as a result of error correction.

Model	χ^2/df	GFI	AGFI	CFI	TLI	NFI	IFI	RMSEA	RMR
Before Error Correction	5.953	0.754	0.734	0.884	0.880	0.868	0.984	0.052	0.065
After Error Correction	4.173	0.907	0.838	0.900	0.903	0.912	0.000	0.046	0.058

After implementing the necessary modifications, the updated model (Fig. **2**) displayed significantly improved fit indices, providing a better data representation. As shown in Table **9**, the post-modification fit indices reflect a more robust alignment with the theoretical framework.

Model Fit Summary

Table **9** shows the initial model after modifications; the result provides a better level of acceptance in fit indices.

The measurement model was validated using structural equation modelling (SEM) *via* Lisrel software. Table **10** summarizes the reliability and validity metrics of the constructs, including Composite Reliability (CR), Average Variance Extracted (AVE), Maximum Shared Variance (MSV), and Maximum Reliability (MaxR). All CR values exceeded the recommended threshold of 0.7, with "Skill and Qualifications" achieving the highest reliability at 0.926. Similarly, AVE values for all constructs were above 0.5, confirming convergent validity (Hair *et al.*, 2011). Notably, "Personality and Work Style" exhibited the highest AVE of 0.804, while the discriminant validity of the constructs was verified as MSV values remained below AVE. These findings confirm the robustness of the measurement model and its suitability for further analysis. The structural equation model (Fig. **3**) encapsulates the validated relationships and hypotheses tested in this study. It demonstrates the supported and unsupported hypotheses (as outlined in Table **11**). It highlights the interconnections among skill and qualifications, interests and motivations, personality and work style, values and culture fit, and well-being.

Table 10. Reliability and validity assessment of constructs using Structural Equation Modelling (SEM).

-	CR	AVE	MSV	MaxR(H)	1	2	3	4	5
Skill and Qualifications	0.926	0.759	0.075	0.952	**0.871**	-	-	-	-
well-being	0.875	0.644	0.002	0.972	-0.047	**0.802**	-	-	-
Interests and Motivations	0.867	0.628	0.003	0.929	-0.028	-0.046	**0.793**	-	-
personality and work style	0.925	0.804	0.075	0.950	0.274***	0.029	-0.053	**0.897**	-
values and Culture Fit	0.878	0.706	0.002	0.880	-0.028	0.037	0.050	0.008	**0.840**

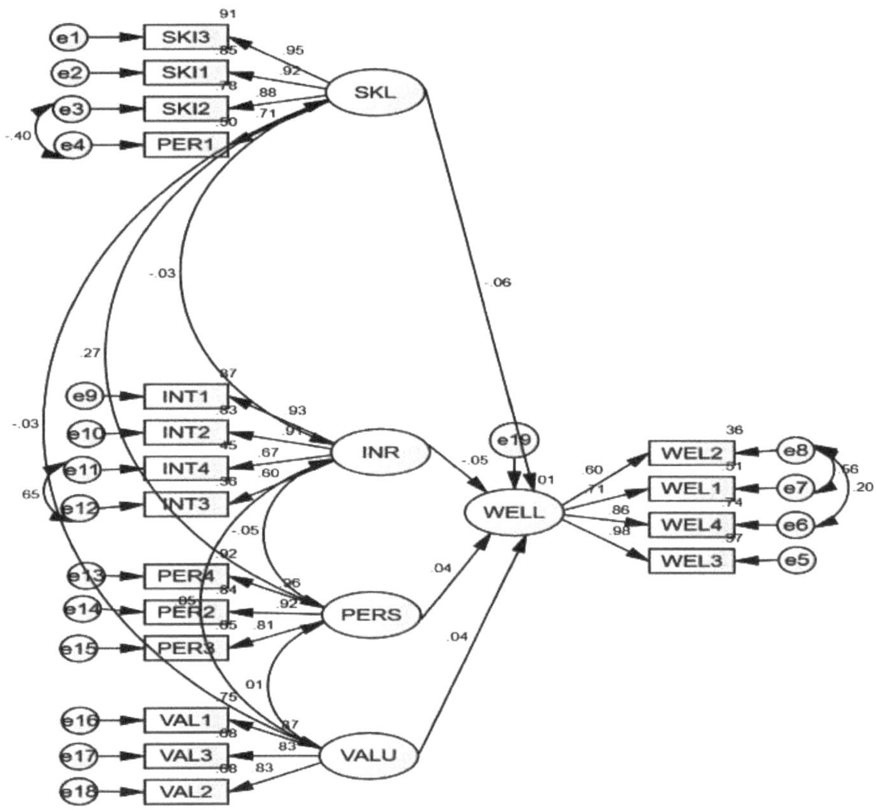

Fig. (3). Structural equation modeling.

Table 11. Hypothesis testing and regression weights.

-	-	-	Hypothesis	p Values	Results
well-being	<---	Skill and Qualifications	H1	.010	Not Supported
well-being	<---	Interests and Motivations	H2	.297	Supported
well-being	<---	personality and work style	H3	.371	Supported
well-being	<---	values and Culture Fit	H4	.015	Not Supported

Regression Weights: (GrouP Number 1 - Default Model)

After verifying measurement characteristics, reliability, and discriminate and convergent validity, we evaluated hypotheses. There was no evidence of a relationship between skill and qualifications and wellbeing ($p = 0.010$, $p < .050$). H1. There was support for the path coefficient of the relationship between well-being and Interests and Motivations ($p = 0.278$, $p > .005$); H2. There was support for the relationship between well-being and personality and work style ($p = 0.371$, p >.005); H3. There was no support for the path coefficient of the relationship between well-being and values and Culture Fit (p = 0.015, p <.005); and H4.

Below are the Policy Recommendations for Enhancing Work-Life Balance and Mental Health in IT Workplaces.

- Mandatory Work-Life Balance Policies IT firms should establish formal policies that restrict after-hours communication and promote the "right to disconnect." Flexitime and compressed workweeks are especially valuable for women balancing work and caregiving responsibilities (Chitra & Ramesh, 2024).
- Formalized Mental Health Support Structures Companies must provide Employee Assistance Programs (EAPs), dedicated mental health leave, and training for managers in Mental Health First Aid. These initiatives have shown measurable benefits in stress reduction among IT employees (World Health Organization, 2022).
- AI-Powered Early Stress Detection Integrating AI tools to monitor email sentiment or biometric feedback can help HR departments identify burnout risks in real time. This is crucial in IT environments where digital overload is common (Deloitte, 2023).
- Redesign Performance Metrics Firms should shift from purely output-based KPIs to well-being-integrated performance goals. This includes rewarding healthy work habits and recognizing often-overlooked emotional labor (Chitra and Ramesh, 2024).
- Caregiver and Childcare Support On-site childcare, dependent care assistance, and return-to-work programs are essential in retaining mid-career women. These benefits significantly improve retention and morale in tech sectors (NASSCOM, 2023).
- Transparent Career Advancement Establish mentorship networks for women and conduct bias audits in promotion cycles. AI tools can track disparities in recognition and leadership access (McKinsey and Company, 2022).
- Wellness-Integrated Learning and Development Blending technical upskilling with well-being training (*e.g.*, mindfulness, time management) empowers employees to sustain high performance without burnout (WHO, 2022).

CONCLUSION

The information technology sector is beset with long work hours, deadlines, and extremely demanding and never-ending work. The specific needs of millennial women, who make up a sizable section of the workforce in the information technology industry, are that their employers should take the initiative to address their demands in order to achieve work-life balance. Without a doubt, millennial women workers have very high aspirations for their careers. However, they also recognize the professional costs associated with changing life stages, such as marriage and parenthood, since these changes also affect their needs and responsibilities. The current study found that millennial women employed in this field struggle to unwind after work because they are too exhausted. It also showed that because work has become so demanding, it is difficult for them to find time to spend with friends and family or engage in their passions. These millennial women professionals find it challenging to attain work-life compatibility since their family duties also interfere with their work-life balance. Since millennials view work as an integral part of their lives, they like to live active lives outside of work and are willing to change careers if their current position does not allow them to attain work-life balance. Enabling millennial women workers to achieve work-life balance is, therefore, a strong competitive advantage for companies in this industry today.

REFERENCES

Arora, R., Vally, D.P. (2018). The impact of work-life balance of workers in information technology companies in India. *IJMEI, 4*(9), 1-8.

Aelenei, C., Martinot, D., Sicard, A., Darnon, C. (2020). When an academic culture based on self-enhancement values undermines female students' sense of belonging, self-efficacy, and academic choices. *J. Soc. Psychol., 160*(3), 373-389.
[http://dx.doi.org/10.1080/00224545.2019.1675576] [PMID: 31600124]

Chauhan, J., Mishra, G., Bhakri, S. (2022). Career success of women: Role of family responsibilities, mentoring, and perceived organizational support. *Vision: The Journal of Business Perspective, 26*(1), 105-117.
[http://dx.doi.org/10.1177/09722629211024887]

Chitra, D., Ramesh, R. (2024). *Examining the role of IT industries towards work compatibility and wellbeing of women employees in Chennai City.*

Croft, A., Schmader, T., Block, K. (2015). An underexamined inequality: cultural and psychological barriers to men's engagement with communal roles. *Pers. Soc. Psychol. Rev., 19*(4), 343-370.
[http://dx.doi.org/10.1177/1088868314564789] [PMID: 25576312]

Deloitte. (2023). 2023 Global human capital trends: The boundaryless workforce. https://www2.deloitte.com.

Geetha, M., Ashtikar, S.P. (2022). Work-life balance among women in the private higher education industry during COVID-19: A path to organizational sustainability. *J. Bus. Res., 57*(3), 455-468.
[http://dx.doi.org/10.1002/jbr.17942]

Heikkinen, S., Kivijärvi, M. (2022). Gendered work–life ideologies among IT professionals. *NORA, 30*(2), 108-123.
[http://dx.doi.org/10.1080/08038740.2021.2015433]

Joseph, J. (2022). Exploring the relationship between personality and work-life balance. *Integrating new technologies in international business.* Apple Academic Press.
[http://dx.doi.org/10.1201/9781003130352-4]

Krishna Kumar. (2018). Conceptual model on the impact of Generation Y employees on information technology organizations. *Journal of Advanced Research in Dynamical & Control Systems, 10*(2), 785-792.

Pyrkosz-Pacyna, J., Dukala, K., Kosakowska-Berezecka, N. (2022). Perception of work in the IT sector among men and women—A comparison between IT students and IT professionals. *Front. Psychol., 13*, 944377.

[http://dx.doi.org/10.3389/fpsyg.2022.944377] [PMID: 36160560]

McKinsey & Company. (2022). Women in the workplace 2022. https://www.mckinsey.com.

NASSCOM. (2023). Inclusion matters: Gender parity in tech workplaces. https://nasscom.in.

World Health Organization. (2022). Mental health at work: Policy brief. https://www.who.int.

CHAPTER 9

Redefining Recruitment: The Role of AI and Automation in Transforming Candidate Selection

Dipti Malpani[1,*], **Supriya Pathak**[1] and **Dheeraj Nim**[1]

[1] *Faculty of Management, Oriental University, Indore, India*

Abstract: The recruitment landscape is undergoing a significant transformation with the integration of Artificial Intelligence (AI) and automation. These technologies streamline traditional recruitment processes such as sourcing, screening, interviewing, and decision-making, significantly enhancing efficiency and reducing time-to-hire. AI-powered tools, like applicant tracking systems and chatbots, offer data-driven insights, minimize biases, and improve candidate experiences. Automation optimizes repetitive tasks, allowing recruiters to focus on strategic priorities. However, the integration of AI also presents challenges, including algorithmic bias, privacy concerns, and the need for ethical practices. This chapter explores the evolution of recruitment from manual methods to AI-driven approaches, highlighting the advantages, challenges, and ethical considerations of using AI in hiring. It delves into practical applications, including predictive analytics, video interviewing, and diversity hiring, while showcasing case studies from organizations leveraging these innovations. The findings underscore AI's potential to revolutionize recruitment and emphasize the importance of balanced human-AI collaboration for effective talent acquisition.

Keywords: AI in recruitment, Automation, Bias reduction, Candidate selection, Ethical practices, Predictive analytics, Talent acquisition.

INTRODUCTION

Recruitment is the process of identifying sources from which candidates can be sourced and brought into the organization. It involves finding a pool of skilled and qualified individuals who are suitable for the organization's needs. “Recruitment is the process of searching for prospective employees and stimulating them to apply for jobs in the organization” (Edwin B. Flippo,1980). As a fundamental function of HR, recruitment plays a crucial role in placing the right candidate in the right job. It is a pivotal activity that directly contributes to achieving the organization's goals. When recruitment is done correctly, it ensures effective sel-

* **Corresponding author Dipti Malpani:** Faculty of Management, Oriental University, Indore, India; E-mails: Diptimaplani@oriental university.in, Diptimalpani84@gmail.com

Rishikaysh Kaakandikar, Keshav Kaushik, Supriya Pathakm Madhavi Shamkuwar & Vivek Mishra (Eds.)

ection, training, and development of employees, which in turn enhances the organization's workforce and drives the timely achievement of its objectives. On the other hand, if recruitment is not handled properly, it leads to a cycle of poor decisions and inefficiencies, commonly referred to as "Garbage in, Garbage out."

This concept is often represented by the **1-10-100 rule**, which highlights that if mistakes are not corrected in the early stages of the recruitment process, the cost to fix them increases exponentially as time goes on. Therefore, taking the right steps at the initial stage of recruitment is crucial and ultimately more cost-effective for the organization.

Recruitment can be a lengthy and complex task for HR departments, involving various sub-processes such as:

Each step of the recruitment process is crucial, but often tedious and time-consuming. Traditional recruitment methods demand significant time, effort, and financial resources. However, these challenges have been addressed with the integration of Artificial Intelligence (AI), which is streamlining the entire process, saving both time and costs while improving efficiency.

Artificial Intelligence (AI) involves simulating human intelligence in machines designed to think, learn, and perform tasks typically requiring human cognitive abilities. These systems analyze data, recognize patterns, make decisions, and solve problems, continuously enhancing their performance through machine learning algorithms. As John McCarthy stated in 1956, "Artificial Intelligence is the science and engineering of making intelligent machines, especially intelligent computer programs." While AI aims to understand and replicate aspects of human intelligence, it is not limited to biologically observable methods.

In essence, AI empowers machines to adapt intelligently to new situations, process complex information, and exhibit behaviors that replicate human thought processes, such as understanding language, recognizing images, and making data-driven predictions.

The recruitment industry is experiencing a significant transformation, largely driven by the adoption of AI and automation technologies, Geetha, R., and Bhanu, S. R. D. (2018). These innovations are reshaping traditional recruitment practices by simplifying processes, enhancing decision-making, and improving the overall candidate experience. This chapter delves into how AI and automation are revolutionizing the recruitment landscape, their impact on candidate selection, and the various challenges and opportunities they bring for organizations.

THE EVOLUTION OF RECRUITMENT THROUGH AI AND AUTOMATION

Recruitment has historically been a labour-intensive process, relying on manual methods for sourcing, screening, and hiring candidates. With the advent of AI and automation, these tasks have become faster, more accurate, and less resource-intensive. AI-powered tools, such as Applicant Tracking Systems (ATS) and resume parsers, automate repetitive tasks, allowing recruiters to focus on strategic decision-making (Eubanks, 2020).

For instance, automation streamlines the initial screening process by analyzing resumes for specific keywords, qualifications, and experience, significantly reducing the time required to identify suitable candidates. AI algorithms use Natural Language Processing (NLP) to assess resumes and cover letters, ensuring a consistent evaluation free from human biases (Upadhyay and Khandelwal, 2018).

Automation and AI have changed the paradigm of Recruitment. Various sub-processes of Recruitment have been made easier with the help of AI tools like:

Applicant Tracking System (ATS): Described as a goldmine of recruitment data, the ATS is portrayed as a hiring assistant capturing candidate applications, CVs, recruitment funnel details, sources, selections, and more.

Optical Character Recognition (OCR): OCR stands out by converting printed and handwritten text into machine-encoded text, presenting a wealth of data from resumes, application forms, and handwritten documents.

Optical Mark Recognition (OMR): OMR is pivotal in simplifying data collection and analysis, especially in recruitment surveys and assessments where respondents mark their choices. This source contributes data on employee preferences and feedback.

Intelligent Character Recognition (ICR): Going beyond printed text recognition, ICR intelligently interprets handwritten text. It transforms handwritten feedback from performance appraisals or training evaluations into valuable data for in-depth analysis.

Intelligent Document Recognition (IDR): IDR is a digital HR assistant that automatically identifies and categorizes documents based on content. This source is significant in providing structured data from resumes, invoices, and various HR records.

QR Code Recognition: QR codes introduce a user-friendly dimension to HR, especially in employee onboarding. Scanning QR codes provides instant access to training materials, policy documents, or benefits information, generating new data sources related to employee engagement and compliance.

Voice Recognition: Voice Recognition emerges as an inclusive data source where individuals use voice commands to access information, submit requests, or provide feedback through voice surveys. This technology enhances user-friendliness and contributes valuable data on employee preferences and concerns.

Chatbots: A chatbot is an Artificial Intelligence (AI)-powered software application designed to simulate human conversation. It can interact with users through text or voice, answering questions, providing information, and performing tasks. Chatbots are widely used in recruitment processes nowadays, Albert, E. T. (2019).

Screening Software: Depending on the context, "Screening software" can refer to various tools designed to evaluate, filter, or analyze data, people, or systems.

Video Interviewing: Video interviewing involves using technology to conduct interviews remotely *via* video communication platforms. It can be live (real-time) or asynchronous (pre-recorded responses), which is widely used in recruitment processes nowadays (Albassam, W. A., 2023).

Social media screening: Social Media Screening analyzes a candidate's or individual's social media profiles to assess their online behavior, character, and alignment with an organization's values.

The evolution of recruitment methods from print to digital has significantly transformed the hiring landscape. Initially, Recruitment was a local process heavily reliant on physical advertisements and personal recommendations (Parry & Tyson, 2008). The 1990s saw the emergence of online job boards, which revolutionized the industry by enabling global job advertisement reach. These platforms gave organizations a broader audience for their vacancies and gave job seekers easier access to diverse opportunities, shifting the focus toward a more candidate-centric approach (Breaugh, 2020).

In recent years, social media platforms and professional networks like LinkedIn have become essential tools in recruitment. They offer advanced targeting capabilities, allowing recruiters to connect with candidates based on their skills, experience, and professional networks. These platforms also enable engagement with passive candidates who may not be actively job hunting, thus improving the overall quality of the talent pool (Koch, Gerber, & De Klerk, 2018).

Furthermore, AI-driven tools now play a crucial role in modern Recruitment, aiding in candidate sourcing, resume screening, and preliminary assessments. These innovations have streamlined the recruitment process, reduced the time to hire, and increased efficiency (Black and van Esch, 2020). This shift from manual methods to AI-powered solutions highlights the continuous evolution of recruitment practices

Traditional Recruitment (Pre-1980s)

Recruitment in the earlier stages heavily relied on **manual and face-to-face methods:**

- **Newspapers and Magazines:** Employers post job advertisements in newspapers, magazines, or job boards to attract candidates.
- **Walk-in Applications:** Candidates would directly approach companies with physical resumes.
- **Employee Referrals:** Hiring through recommendations from current employees.
- **Job Fairs and Events:** Employers attended physical fairs to source talent.
- **Internal Recruitment:** Companies prefer hiring internally for promotions and new roles.
- **Hiring Agencies:** Specialized firms helped companies find talent through manual search processes.

Limitations

- Time-consuming, expensive, and localized.
- Limited access to a diverse talent pool.
- Lack of data for evaluating hiring decisions.

The Advent of Technology (1980s–1990s)

With the rise of **computers and the internet**, recruitment saw significant changes:

- **Job Portals:** Websites like Monster (est. 1994) and CareerBuilder revolutionized online job advertising.
- **Email Applications:** Candidates could send digital resumes *via* email, reducing paper dependency.
- **Database Management:** Recruiters started using software to store and search resumes more efficiently.
- **Applicant Tracking Systems (ATS):** Early ATS allowed employers to manage applications systematically.

Impact

- Recruitment became more organized and faster.
- Employers reached a broader audience beyond their local areas.
- However, it was still dependent on manual human judgment for decision-making.

Social Media and Digital Era (2000s–2010s)

The **rise of social media** and digital tools created a new paradigm in recruitment:

- **LinkedIn (2003):** Enabled recruiters to search for passive and active candidates globally.
- **Social Media Platforms:** Facebook, Twitter, and other platforms were used for job postings and employer branding.
- **Online Assessment Tools:** Recruiters adopted tools to conduct skill assessments, psychometric tests, and video interviews.
- **Search Algorithms:** Technology-driven job matching began with keyword searches and skills-based filtering.

Advantages

- Faster outreach, broader candidate pool, and better employer branding.
- Improved candidate experience through digital processes.
- Data-driven decision-making gained traction.

Challenges

- Resume keyword stuffing became prevalent to bypass filtering.
- High volumes of irrelevant applicants required more screening.

AI-Driven Recruitment (2015–Present)

The introduction of **Artificial Intelligence (AI)** transformed recruitment by automating and optimizing processes:

Key AI Tools and Features

AI-Powered ATS:

- AI automates resume screening by analyzing keywords, skills, and experiences.
- Reduces recruiter bias through objective decision-making.

Chatbots

- AI chatbots like **Mya** or **Paradox** answer candidate queries and assist with the application process 24/7.

Candidate Sourcing Tools

- AI tools proactively search for candidates on LinkedIn, job boards, and social media platforms.

Predictive Analytics

- AI predicts candidate success based on historical data and trends, helping recruiters make informed decisions.

Video Interview AI

- During interviews, tools like **HireVue** analyze candidates' body language, tone, and word choice.

Skill Assessments

- AI-based testing platforms provide unbiased assessments of candidates' skills, reducing hiring errors.

Diversity Hiring

- AI eliminates unconscious biases by focusing on data and skill matches rather than demographics.

Benefits of AI in Recruitment

- **Speed:** Significantly reduces time-to-hire.
- **Efficiency:** Automates repetitive tasks like screening and scheduling.
- **Cost Reduction:** Minimizes hiring expenses.
- **Better Candidate Experience:** AI tools provide timely feedback and personalized communication.
- **Improved Quality of Hire:** Data-driven insights enhance hiring accuracy.

Challenges

- **Bias in AI:** If AI systems are trained on biased data, they can perpetuate biases.
- **Human Touch:** Over-reliance on AI can reduce personalization and emotional connection.
- **Technical Limitations:** Small organizations may struggle with costs and implementation.

Recruitment has evolved from **manual, paper-based processes** to a highly automated, AI-driven system. While AI has revolutionized efficiency, speed, and decision-making, organizations still strive to balance **human judgment** and **technological tools** to ensure effective and inclusive hiring.

APPLICATIONS OF AI AND AUTOMATION IN CANDIDATE SELECTION

Various functions of recruitment can be broadly divided into four parts:

The findings indicate that AI in recruitment is relatively new, with few companies fully integrating AI into every aspect of their recruitment processes. The most effective applications of AI in traditional recruitment are in activities like pre-selection, candidate communication, and delivering recruitment results to applicants. Key benefits of AI include increased efficiency, improved quality, and the automation of routine tasks. However, companies' readiness to adopt and implement new technologies is a significant challenge (Johansson, J., & Herranen, S., 2019).

Sourcing Candidates

AI tools have revolutionized the sourcing process by leveraging big data to identify and attract potential candidates. Platforms like LinkedIn and other recruitment databases use AI to recommend candidates based on job descriptions and hiring managers' preferences. AI-powered chatbots also engage with passive candidates, answering queries and providing information about job openings, thus expanding the talent pool (Chamorro-Premuzic *et al.*, 2019).

For example, predictive analytics identifies individuals likely to switch jobs based on online activity, enabling proactive recruitment strategies. These capabilities improve talent acquisition efficiency and ensure a better match between candidates and organizational needs.

Screening and Shortlisting

AI enhances the screening and shortlisting process by analyzing resumes and applications against predefined criteria. Advanced algorithms accurately assess candidates' skills, qualifications, and experience, eliminating unqualified applicants and saving recruiters significant time.

Additionally, AI tools can evaluate candidates' social media profiles, work samples, and professional networks to better understand their suitability for a role. This multi-dimensional analysis ensures that only the most relevant candidates proceed to the next stages of the hiring process (Eubanks, 2020).

Interviewing

AI-driven video interview platforms like HireVue and Pymetrics analyze candidates' responses, facial expressions, and speech patterns to assess personality traits, communication skills, and cultural fit. These platforms use ML algorithms to compare candidates against successful employee profiles, ensuring objective evaluations (Upadhyay & Khandelwal, 2018).

Furthermore, automated scheduling systems reduce logistical challenges, allowing candidates to choose interview slots at their convenience. These systems enhance the candidate experience and demonstrate organizational efficiency.

Assessment and Decision-Making

AI-powered assessments provide insights into candidates' cognitive abilities, technical skills, and emotional intelligence. Gamified tests, psychometric evaluations, and simulations offer data-driven insights into candidates' potential performance.

AI also supports decision-making by predicting candidates' long-term success within an organization. These predictions are based on previous work performance, tenure, and compatibility with company culture (Chamorro-Premuzic *et al.*, 2019).

APPLICATIONS OF AI AND AUTOMATION IN SUPPORT FUNCTIONS OF RECRUITMENT AND SELECTION

AI is transformative in enhancing efficiency, personalization, and effectiveness in HR functions like **employee onboarding**, **engagement**, **relations**, **career development**, and **joining formalities,** Geetha, R., & Bhanu, S. R. D. (2018). Here's how AI is applied to these areas:

Employee Onboarding

AI streamlines the onboarding process, making it faster, more personalized, and engaging for new hires.

Applications

- **Chatbots for FAQs:** AI-powered chatbots (*e.g.*, Talmundo, Enboarder) provide instant answers to new hires' questions about policies, benefits, and processes.
- **Task Automation:** Automates administrative tasks like filling out forms, document submission, and account setup.

- **Personalized Onboarding Plans:** AI tools create customized onboarding schedules based on the employee's role, location, and team.
- **Virtual Assistants:** Platforms like BambooHR guide new hires through training materials and orientation programs.

Benefits

- Reduces administrative workload for HR teams.
- Improves employee experience by providing 24/7 support.
- Speeds up time-to-productivity for new hires.

Employee Engagement

AI fosters engagement by providing actionable insights into employee behaviour and sentiment, enabling tailored interventions.

Applications

- **Sentiment Analysis:** Tools like Workday or Peakon analyze employee feedback from surveys, emails, or chats to measure morale and engagement.
- **Personalized Recommendations:** AI suggests wellness programs, learning resources, or team-building activities based on individual preferences.
- **Virtual Recognition Systems:** AI tools like Kazoo enable managers to recognize employees' achievements and recommend rewards based on performance data.
- **Pulse Surveys:** AI-powered surveys adapt questions dynamically to gather more relevant insights about employee sentiment.

Benefits

- It enhances employee satisfaction and reduces attrition.
- Helps HR identify and address engagement challenges proactively.
- Encourages a culture of recognition and inclusivity.

Employee Relations

- AI improves employee relations management by offering data-driven insights and ensuring fair practices.

Applications

- **Conflict Resolution:** AI chatbots mediate disputes by guiding employees to resources, policies, or appropriate HR contacts.
- **Policy Adherence:** AI tools monitor compliance with workplace policies and flag potential issues like harassment or misconduct.

- **Predictive Analytics:** Tools like IBM Watson predict potential conflicts or attrition risks based on historical patterns and sentiment data.
- **Feedback Analysis:** AI analyzes employee feedback to identify recurring issues and suggest systemic improvements.

Benefits

- Encourages proactive conflict management and resolution.
- Improves transparency and fairness in employee relations.
- Enables HR to act on data-backed insights for better decision-making.

Career Development

AI empowers employees to take charge of their career growth by offering personalized learning and development paths.

Applications

- **Skill Gap Analysis:** AI tools like Eightfold.ai identify skills employees need to progress in their careers and suggest training programs.
- **Mentorship Matching:** AI matches employees with mentors based on skills, goals, and compatibility.
- **Learning Recommendations:** Platforms like Degreed use AI to recommend courses, certifications, and workshops tailored to an employee's career path.
- **Internal Mobility Platforms:** AI suggests suitable internal roles based on an employee's skills, interests, and performance.

Benefits

- Enhances employee retention by fostering growth opportunities.
- Aligns individual career goals with organizational objectives.
- Encourages continuous learning and upskilling.
- Joining Formalities

AI simplifies and accelerates the joining process, ensuring a smooth start for new hires.

Applications

- **Document Verification:** AI tools automatically verify uploaded documents like IDs, certificates, and work permits.
- **Digital Contracts:** Platforms like DocuSign leverage AI to ensure accurate and legally compliant contract generation and signing.
- **Pre-boarding Portals:** AI-powered portals guide candidates through tasks like filling out personal details, setting up accounts, and selecting benefits.

- **Chatbot Assistance:** Virtual assistants help candidates with real-time updates on their application and onboarding progress.

Benefits

- Reduces errors in documentation and compliance.
- Improves communication between new hires and HR teams.
- Minimizes delays in onboarding.

ADVANTAGES OF AI AND AUTOMATION IN RECRUITMENT

Efficiency and Speed

AI and automation drastically reduce the time-to-hire by handling repetitive tasks such as resume screening, interview scheduling, and candidate communication. This efficiency enables recruiters to fill positions faster, minimizing disruptions to business operations (Upadhyay & Khandelwal, 2018).

Enhanced Candidate Experience

Automated communication tools, such as chatbots and personalized emails, keep candidates informed and engaged throughout the hiring process. A seamless and transparent recruitment experience improves candidates' perception of the organization, enhancing employer branding (Eubanks, 2020).

Bias Reduction

When designed and implemented ethically, AI algorithms reduce unconscious biases in candidate evaluation. AI ensures fairer hiring practices and promotes workplace diversity by focusing on objective criteria, such as skills and qualifications (Chamorro-Premuzic *et al.*, 2019).

Cost Savings

Automation reduces administrative costs associated with recruitment by streamlining processes and optimizing resource allocation. Organizations can achieve better outcomes with fewer resources, increasing the overall Return On Investment (ROI).

CHALLENGES OF AI AND AUTOMATION IN RECRUITMENT

Algorithmic Bias

While AI can potentially reduce human bias, it is not immune to bias. If the training data used to develop AI systems reflects existing biases, these biases can be perpetuated in hiring decisions. For instance, an AI system trained on historical hiring data may inadvertently favour candidates from specific demographics (Upadhyay & Khandelwal, 2018).

Data Privacy Concerns

Recruitment processes involve collecting and processing sensitive candidate data. Organizations must comply with data privacy regulations, such as the General Data Protection Regulation (GDPR), to ensure the ethical use of AI in recruitment (Eubanks, 2020).

Over-Reliance on Technology

Excessive reliance on AI and automation may result in a lack of human oversight and intuition in hiring decisions. While AI excels at processing data, it cannot replicate the nuanced judgment of experienced recruiters. Balancing automation with human involvement is essential for effective recruitment.

Adoption and Integration Challenges

Implementing AI-driven recruitment tools requires significant investment in technology and training. Organizations must also ensure seamless integration with existing systems and processes to maximize the benefits of these technologies (Chamorro-Premuzic *et al.*, 2019).

Lack of Transparency in AI Decision-Making

One significant challenge with AI in recruitment is the lack of transparency in decision-making. Many AI systems operate as "black boxes," where the logic behind their recommendations is difficult to understand or explain. This opacity can make it challenging for recruiters to trust the outcomes and for organizations to ensure compliance with fairness and ethical standards (Zhang *et al.*, 2020). Transparent AI systems that provide explainable results are essential to build trust and accountability in recruitment.

Resistance to Change among Stakeholders

Another challenge is the resistance to adopting AI and automation among employees and hiring managers. Many stakeholders may fear AI will replace their roles or doubt its ability to make accurate hiring decisions. Overcoming this resistance requires effective communication, training programs, and a demonstration of the value of technology in complementing human efforts rather than replacing them (Huang & Rust, 2021).

ETHICAL ISSUES WITH AI AND AUTOMATION IN RECRUITMENT

AI and automation in recruitment present several ethical issues that need careful consideration. These challenges revolve around fairness, transparency, privacy, and accountability Rigotti, C., & Fosch-Villaronga, E. (2024). Below are the key ethical concerns:

Bias and Discrimination

Reinforcement of Biases: AI systems often learn from historical data, which may include biases related to gender, race, age, or socioeconomic status. This can perpetuate and even amplify discrimination.

Unintentional Discrimination: Automated systems might screen out qualified candidates because of data patterns that correlate with irrelevant factors.

Lack of Transparency

- **Black Box Algorithms:** Many AI systems operate as "black boxes," making it difficult to understand how decisions are made. This lack of transparency can erode trust and make it hard for candidates to contest decisions.
- **Unclear Criteria:** Job applicants may not know what factors are being considered in the evaluation process.

Privacy Concerns

- **Data Collection:** Automated systems often rely on large amounts of personal data, raising concerns about how this data is collected, stored, and used.
- **Invasive Profiling:** AI tools that analyze social media activity or other personal data to assess candidates can intrude on privacy.

Accountability and Governance

- **Responsibility for Errors:** When an AI system makes a mistake—such as rejecting a qualified candidate—it is often unclear who is responsible: the developer, the employer, or the AI itself.
- **Ethical Oversight:** Many organizations lack adequate oversight mechanisms to ensure the ethical use of AI in recruitment.

Fair Access to Opportunities

- **Digital Divide:** Automated recruitment processes may disadvantage candidates without access to technology or those unfamiliar with digital platforms.
- **Over-reliance on Automation:** Relying solely on AI might exclude candidates who excel in soft skills that algorithms do not measure easily.

Dehumanization of Recruitment

- **Reduced Human Interaction:** Automation can make hiring impersonal, leaving candidates undervalued or overlooked.
- **Cultural Fit:** AI might struggle to evaluate nuanced factors like a candidate's cultural fit or interpersonal skills.

Manipulation and Gaming

- **Algorithmic Manipulation:** Savvy candidates might learn to game AI systems by tailoring their applications to keywords or patterns favoured by the algorithms.
- **Ethical Implications of Tools:** Tools like resume scanners can encourage over-optimization, favouring quantity over quality in applications.

ADDRESSING ETHICAL ISSUES

Organizations can take several steps to address these ethical concerns:

- **Bias Audits:** Regularly evaluate AI systems for biases and implement corrective measures.
- **Transparency:** Clearly communicate how AI systems work and their decision-making criteria.
- **Privacy Safeguards:** Establish robust data protection policies and obtain candidate consent for data use.
- **Human Oversight:** Ensure final hiring decisions involve human judgment, not solely automated systems.
- **Regulatory Compliance:** Adhere to laws and ethical standards governing AI and employment practices.

THE FUTURE OF AI AND AUTOMATION IN RECRUITMENT

As AI and automation continue to evolve, their role in recruitment is expected to expand further. Emerging technologies, such as Natural Language Generation (NLG) and advanced analytics, will enhance candidate engagement and provide deeper insights into talent acquisition trends.

For example, AI-powered tools may predict future workforce needs based on market trends and organizational growth patterns, enabling proactive recruitment strategies. Additionally, real-time analytics will help organizations measure the effectiveness of their hiring processes, driving continuous improvement (Upadhyay & Khandelwal, 2018).

The ethical use of AI in recruitment will also gain prominence, with organizations prioritizing transparency, accountability, and fairness in their AI systems. Collaboration between technologists, HR professionals, and policymakers will be crucial to addressing the ethical and regulatory challenges associated with AI-driven recruitment.

PRACTICAL INSIGHTS INTO THE APPLICATIONS OF AI AND AUTOMATION IN RECRUITMENT

HireVue and AI-Powered Video Interviews

HireVue serves as a prominent example of AI-driven recruitment. This platform is widely utilized for conducting video interviews within the hiring process, employing AI technology to assess candidates based on their responses in pre-recorded or live video interviews. Headquartered in South Jordan, Utah, HireVue is an artificial intelligence and human resources management company founded in 2004. It enables clients to conduct digital interviews, allowing candidates to interact with a computer rather than a human interviewer. Through its AI capabilities, HireVue analyzes video, audio, and text data from these interviews to evaluate candidates effectively. The AI assesses several factors:

- **Verbal communication:** Tone, vocabulary, and fluency.
- **Nonverbal cues:** Facial expressions and body language (in earlier versions, this aspect has been criticized and may have been revised).
- **Content analysis:** The relevance and quality of a candidate's answers.

These insights are scored and provided to recruiters to help them make informed decisions.

Kaggle and GitHub Datasets on Resume Screening

Kaggle and GitHub are excellent resources for finding datasets on **resume screening** and related topics. These datasets can be used to train machine learning models, conduct experiments, or explore trends in recruitment.

Datasets Available on Kaggle

Kaggle hosts various datasets specifically curated for machine learning and data science projects, including resume and recruitment data.

Popular Datasets for Resume Screening

Resume Dataset

- **Description:** Includes resumes categorized by skills, experience, and job roles and often used to build or evaluate resume-parsing tools or matching systems.
- **Applications:** Natural Language Processing (NLP) for keyword extraction, skill-matching algorithms, and AI-based candidate screening.
- Example on Kaggle: Search for "resume" or "HR analytics."

Job Descriptions and Resumes

- **Description:** Datasets containing job descriptions and resumes to train matching algorithms.
- **Applications:** Compare resumes with job descriptions to rank candidates for a position.

HR Analytics Datasets

- **Description:** Data on recruitment pipelines, including applicant tracking and hiring outcomes.
- **Applications:** Analyzing hiring trends, optimizing the hiring funnel, and identifying bottlenecks.

Text Classification Datasets

- **Description:** Datasets used for classifying resumes by domain, role, or skillset.
- **Applications:** NLP techniques to categorize resumes automatically.

Datasets Available on GitHub

GitHub is a goldmine for open-source datasets and projects, especially when used in conjunction with tools for resume parsing and analysis.

Popular Repositories for Resume Screening

- **Resume Parsing Tools with Sample Data**
  - Many repositories include datasets alongside code for building parsers or matching systems.
  - **Example Repos:**
  - Tools like RESUME-PARSER or PyRE Sparser with annotated datasets.
  - Resume-to-job-matching ML models.

Synthetic Resume Datasets

- Repositories hosting synthetic resumes for research or training, which mimic real-world patterns while avoiding privacy issues.
- Applications: Model training and benchmarking.

Job Matching Projects

- Complete projects with job description datasets and resumes to train ranking or recommendation systems.
- **Example:** Projects tagged with "NLP," "resume-matching," or "HR analytics."

LinkedIn Recruiter

- Features AI-based recommendations for potential candidates based on job descriptions and recruiter preferences.

CASE STUDY

Case studies demonstrate how AI and automation have transformed recruitment processes across industries, highlighting their impact, challenges, and outcomes.

Unilever – AI for High-Volume Recruitment

Industry: Consumer Goods

Problem: Unilever faced challenges managing high application volumes (250,000+ annually) while ensuring a positive candidate experience.

Solution

- Partnered with **HireVue** to automate the early stages of recruitment.
- Implemented:
- **AI-powered video interviews:** Candidates answered preset questions, and AI evaluated their responses based on communication skills and behavioral traits.
- **Gamified assessments:** Pymetrics games assessed cognitive and emotional traits.
- **Automated resume screening:** AI matched resumes to job requirements.

Impact

- Reduced recruitment time by **75%** (from four months to two weeks).
- Improved diversity and inclusion by anonymizing candidate data during the screening phase.
- Increased candidate satisfaction, with **92% positive feedback** on the process.

Hilton Hotels – Automating Recruitment for Seasonal Hires

Industry: Hospitality

Problem: Hilton needed to fill seasonal roles quickly, often hiring thousands of employees within a short timeframe.

Solution

- Used **Paradox Olivia**, an AI-powered chatbot, to:
- Automate candidate engagement *via* chat.
- Pre-screen applicants by asking role-specific questions.
- Schedule interviews by syncing with recruiters' calendars.

Impact

- Saved over **100,000 recruiter hours** annually.
- Reduced time-to-hire from 6 weeks to **5 days**.
- Increased application completion rates by **40%**, as the chatbot made the process more intuitive.

Vodafone – AI to Improve Diversity in Hiring

Industry: Telecommunications

Problem: Vodafone wanted to address unconscious bias in its recruitment process and increase diversity in its workforce.

Solution

- Partnered with **Applied**, an AI tool designed to anonymize applications and focus on skills-based assessments.
- Implemented:
- **Anonymized applications:** Removed identifying details like names and education history.
- **AI-driven matching:** Ranked candidates based on skill alignment rather than experience or demographics.

Impact

- Increased female representation in technical roles by **20%**.
- Improved diversity across multiple dimensions (*e.g.*, ethnicity, socioeconomic background).
- Enhanced recruiter trust in the system due to transparent scoring and ranking.

Amazon – Scaling High-Volume Hiring with AI

Industry: E-commerce

Problem: Amazon needed to hire hundreds of thousands of warehouse workers during peak seasons, such as Black Friday and the holiday season.

Solution

- Leveraged **AI-driven screening tools** to analyze applications quickly.
- Used chatbots to engage with candidates and schedule interviews.
- Implemented predictive analytics to forecast candidate retention and performance.

Impact

- Reduced time-to-hire by **30%**.
- Achieved scalability, hiring **100,000+ workers** in a matter of weeks.
- Enhanced operational efficiency by focusing on candidates likely to perform well and stay longer.

L'Oréal – AI for Global Recruitment

Industry: Beauty and Personal Care

Problem: L'Oréal faced inefficiencies in managing global recruitment processes across 145 countries.

Solution

- Introduced an AI-powered recruitment assistant:
- **Chatbot (Mya)** to engage with candidates, answer FAQs, and schedule interviews.
- Automated resume screening to identify top talent faster.
- AI tools to analyze candidate engagement and recommend personalized follow-ups.

Impact

- Reduced recruitment time by **70%**.
- Enhanced candidate engagement, with 92% of candidates completing the chatbot interaction.
- Saved recruiters hundreds of hours, allowing them to focus on strategic tasks.

IBM – AI for Internal Mobility

Industry: Technology

Problem: IBM needed to address internal skills gaps and reduce hiring costs by promoting internal mobility.

Solution

- Developed **Watson Talent**, an AI-powered tool to:
- Match internal candidates to roles based on their skills, experience, and career goals.
- Identify skill gaps and recommend training opportunities.
- Predict employee success in new roles using historical data.

Impact

- Filled **40% of roles internally**, reducing external hiring costs.
- Increased employee retention by providing clear career growth pathways.

Enhanced employee satisfaction by aligning roles with personal development goals.

ETHICAL SAFEGUARDS AND BIAS MITIGATION STRATEGIES IN AI-DRIVEN CANDIDATE SCREENING

Bias Auditing and Debiasing Algorithms

Regular audits of AI recruitment systems are essential to detect bias across gender, race, and age. Tools such as IBM AI Fairness 360 and Aequitas enable organizations to measure and mitigate discriminatory patterns in machine learning models (Bellamy *et al.*, 2018). Bias mitigation should be implemented during the training phase through fairness constraints and post-processing evaluations.

Explainable AI (XAI) for Transparency

To enhance trust and accountability, companies should adopt explainable AI techniques. These models provide interpretable outputs, enabling HR professionals and candidates to understand the reasoning behind hiring decisions (Rai, 2020). Explainability also aids in contesting decisions that may appear biased or unfair.

Anonymization of Candidate Data

Removing identifying information such as name, age, gender, and university from applications can reduce unconscious bias (Binns *et al.*, 2018). For instance, Blendoor anonymizes candidate profiles, leading to improved diversity in tech recruitment pipelines.

Human-in-the-Loop Oversight

AI systems should support—not replace—human decision-making. Recruiters must be trained to evaluate AI recommendations critically and intervene when needed. This hybrid approach balances efficiency with ethical responsibility (Binns, 2019).

Diverse and Inclusive Training Data

Training datasets must be representative across demographic lines. One of the most cited failures in AI recruiting was Amazon's AI tool, which developed gender bias due to skewed historical data (Dastin, 2018). Balanced, current datasets prevent such discriminatory behavior from being encoded in algorithms.

Compliance with Ethical AI Guidelines

Organizations should adopt standards such as the EU's Ethics Guidelines for Trustworthy AI, which emphasize human agency, fairness, and technical robustness (European Commission, 2019). These frameworks guide ethical design, deployment, and oversight of AI tools in recruitment.

CONCLUSION

AI and automation are transforming the recruitment landscape, offering significant advantages in efficiency, candidate experience, and decision-making. By automating repetitive tasks, these technologies free recruiters to focus on strategic priorities, enhancing overall recruitment effectiveness. However, organizations must address challenges such as algorithmic bias, data privacy concerns, and over-reliance on technology to unlock the full potential of AI in recruitment.

As the adoption of AI and automation accelerates, their impact on recruitment will only grow. This will shape a future where hiring decisions are data-driven, unbiased, and candidate-centric. Organizations that embrace these technologies while adhering to ethical principles will position themselves as leaders in the competitive talent acquisition landscape.

REFERENCES

Albassam, W.A. (2023). The power of artificial intelligence in recruitment: An analytical review of current AI-based recruitment strategies. *International Journal of Professional Business Review, 8*(6), e02089. [http://dx.doi.org/10.26668/businessreview/2023.v8i6.2089]

Albert, E.T. (2019). AI in talent acquisition: a review of AI-applications used in recruitment and selection. *Strategic HR Rev., 18*(5), 215-221. [http://dx.doi.org/10.1108/SHR-04-2019-0024]

Breaugh, J., Hammerschmid, G., Rackwitz, M., & Palaric, E. (2020). Research report on collaborative management for ICT-enabled public sector innovation: Work package 6–Deliverable 6.4.

Black, J.S., van Esch, P. (2020). AI-enabled recruiting: What is it and how should a manager use it? *Bus. Horiz., 63*(2), 215-226. [http://dx.doi.org/10.1016/j.bushor.2019.12.001]

Bellamy, R.K.E., Dey, K., Hind, M., Hoffman, S.C., Houde, S., Kannan, K. (2018). AI Fairness 360: An extensible toolkit for detecting, understanding, and mitigating unwanted algorithmic bias. *1810.01943.*

Binns, R. (2019). On the apparent conflict between individual and group fairness. *ACM Conference on Fairness, Accountability, and Transparency,* 514-524.

Binns, R., Veale, M., Van Kleek, M., Shadbolt, N. (2018). It's reducing a human being to a percentage': Perceptions of justice in algorithmic decisions. *CHI Conference on Human Factors in Computing Systems,* 1-14. [http://dx.doi.org/10.1145/3173574.3173951]

Chamorro-Premuzic, T., Winsborough, D., Sherman, R.A., Hogan, R. (2019). New talent signals: Shiny new objects or a brave new world? *Ind. Organ. Psychol., 12*(2), 151-168. [http://dx.doi.org/10.1017/iop.2019.8]

Dastin, J. (2018, October 10). Amazon scrapped 'secret AI recruiting tool' that showed bias against women. Reuters. Avialable from: https://www.reuters.com/article/us-amazon-com-jobs-automation-in-ight-idUSKCN1MK08G.

European Commission. (2019). Ethics guidelines for trustworthy AI. Avialable from: https:/ec.europa.eu/futurium/en/ai-alliance-consultation.

Eubanks, B. (2020). *Artificial intelligence for HR: Use AI to support and develop a successful workforce..* Kogan Page Publishers.

Geetha, R., Bhanu, S.R.D. (2018). Recruitment through artificial intelligence: A conceptual study. *International Journal of Mechanical Engineering and Technology, 9*(7), 63-70.

Huang, M.H., Rust, R.T. (2021). A strategic framework for artificial intelligence in marketing. *J. Acad. Mark. Sci., 49*(1), 30-50. [http://dx.doi.org/10.1007/s11747-020-00749-9]

Johansson, J., Herranen, S. (2019). The application of artificial intelligence (AI) in human resource management: Current state of AI and its impact on the traditional recruitment process. *J. Bus. Res., 57*(4), 112-118.

Koch, T., Gerber, C., De Klerk, J.J. (2018). The impact of social media on recruitment: Are you LinkedIn? *SA J. Hum. Resour. Manag., 16*(1), 1-14. [http://dx.doi.org/10.4102/sajhrm.v16i0.861]

Parry, E., Tyson, S. (2008). An analysis of the use and success of online recruitment methods in the UK. *Hum. Resour. Manage. J., 18*(3), 257-274. [http://dx.doi.org/10.1111/j.1748-8583.2008.00070.x]

Rigotti, C., Fosch-Villaronga, E. (2024). Fairness, AI & recruitment. *Comput. Law Secur. Rev., 53*, 105966. [http://dx.doi.org/10.1016/j.clsr.2024.105966]

Rai, A. (2020). Explainable AI: from black box to glass box. *J. Acad. Mark. Sci., 48*(1), 137-141. [http://dx.doi.org/10.1007/s11747-019-00710-5]

Upadhyay, A.K., Khandelwal, K. (2018). Applying artificial intelligence: implications for recruitment. *Strategic HR Rev., 17*(5), 255-258. [http://dx.doi.org/10.1108/SHR-07-2018-0051]

Zhang, B., Dafoe, A., McKelvey, K. (2020). Explainability and accountability in AI decision-making: A multidisciplinary perspective. *AI Soc., 35*(2), 473-486. [http://dx.doi.org/10.1007/s00146-019-00905-4]

CHAPTER 10

The Future of Recruitment: How AI and Automation are Shaping Candidate Selection The Future of Recruitment

Khursheed Jamil[1,*] and **Dharmendra Kumar**[2]

[1] *Unity PG College, University Of Lucknow, U.P., India*

[2] *Faculty of Law, Shia P.G. College, Lucknow, U.P., India*

Abstract: In today's world of recruitment, artificial intelligence has become a game-changer in how we select candidates for jobs. This study dives into how AI-driven ranking models are reshaping the hiring process, exploring their benefits, challenges, and ethical implications. These models, powered by machine learning, help organizations sift through large pools of applicant data more efficiently, making hiring smoother and promoting diversity in the workforce. However, as we embrace these AI tools, we must also confront issues of fairness, bias, privacy, and transparency.

AI-driven ranking models offer many advantages. They automate tasks like resume screening and candidate matching, freeing up recruiters to focus on strategic decisions. Plus, they analyze candidate data to uncover patterns, aiding in data-driven recruitment choices. There's also potential for these models to reduce bias by relying on objective factors rather than human judgment.

Yet, there are hurdles and ethical dilemmas to consider. Algorithmic bias is a major concern, as these models could unintentionally perpetuate inequalities in hiring. Overcoming bias demands careful attention to data quality and ensuring these models are transparent and fair. There are also ethical questions around privacy and data security that need addressing.

Looking ahead, collaboration across disciplines and ethical AI development will shape the future of candidate selection. By fostering dialogue between data experts, ethicists, and HR professionals, we can make AI-driven recruitment more transparent and accountable. Future research should focus on creating fair AI solutions and finding ways to promote diversity while minimizing bias. Through responsible AI use, we can refine candidate selection processes and foster more inclusive workplaces in the digital era.

* **Corresponding author Khursheed Jamil:** Unity PG College, University Of Lucknow, U.P., India; E-mail: khursheed.sheikh@gmail.com, dk535272@gmail.com

Keywords: AI-driven recruitment, Automation, Algorithmic bias, Candidate selection, Data privacy, Ethical AI practices, Machine learning models.

INTRODUCTION

Artificial intelligence has swiftly emerged as a revolutionary influence across various industries, including recruitment. The intricacies of contemporary recruitmentoverseeing extensive applicant pools, enhancing efficiency, and fostering diversityhave compelled organisations to implement more advanced technological tools. AI-driven recruitment, especially *via* machine learning ranking algorithms, has significantly transformed the methods organisations use to identify, assess, and employ individuals. These models are engineered to analyse extensive candidate data and facilitate hiring decisions. Nonetheless, whereas the potential of AI in recruitment is extensive, it also prompts significant concerns around equity, bias, transparency, and ethical accountability.

Prior to the advent of AI, recruitment was an exceedingly human and laborious endeavour. Human recruiters would meticulously examine several resumes, depend on personal networks, and invest considerable time in interviewing applicants. There are also decisions contending with prejudices and subjectivity. The initial phase of the digital revolution in recruiting commenced in the late 1990s and early 2000s with the introduction of applicant tracking systems (Bresciani *et al*, 2021). These systems automated the collection, sorting, and storage of resumes. Applicant Tracking Systems enabled recruiters to sift through resumes using designated keywords, hence expediting the candidate evaluation process. Nevertheless, these initial systems were primitive and could not comprehensively evaluate a candidate's potential beyond basic keyword matching. If an applicant fails to include the specific term sought by a recruiter, their application may be rejected, regardless of their qualifications.

The second wave of technological revolution arose with the advent of data analytics and cloud computing in the late 2000s and early 2010s (Uma *et al*, 2023). Organisations begin utilising extensive data to enhance hiring decisions, examining historical recruitment patterns, employee performance, and market information to forecast future success. Nonetheless, at this juncture, the process continued to be predominantly reactive as recruiters would gather and assess data manually or with minimal automation.

The 2010s heralded the emergence of AI in recruiting (Charlwood and Guenole, 2024). AI-driven solutions are becoming incorporated into multiple phases of the employment process. These technologies commenced the analysis of more intricate, unstructured data, including social media profiles, online exams, and video interviews, employing natural language processing and machine learning techniques (Allal-Chérif *et al.*, 2021). AI advanced recruitment by automating tedious operations and providing intelligent, data-driven forecasts regarding candidate success, cultural compatibility, and long-term performance. Platforms like LinkedIn Recruiter and other AI-driven tools have significantly enhanced the methods by which recruiters identify and evaluate prospects (Pradhan *et al.*, 2023). Furthermore, AI enables the ranking of candidates according to objective criteria, substantially decreasing time-to-hire and offering insights that are beyond the capabilities of conventional systems.

A primary reason for the extensive implementation of AI in recruitment is its capacity to swiftly and accurately analyse substantial datasets. Conventional recruitment methods frequently exhibit inefficiencies, including human bias, protracted delays, and an absence of data-driven insights. AI and machine learning models mitigate these challenges by equipping recruiters with instruments that can impartially assess thousands of candidates in a significantly reduced timeframe compared to human evaluators.

AI methods are progressively utilised to improve diversity in recruitment (Dwivedi *et al.*, 2021). By concentrating on objective data points, such as abilities and experience. AI-driven models can mitigate the impact of unconscious bias that frequently influences human decision-making. Numerous AI recruitment platforms anonymise candidate data, such as names and images, to guarantee that decisions are made solely on qualifications, hence mitigating potential discrimination based on personal attributes.

Notwithstanding these advantages, substantial hurdles persist in AI-driven recruitment, especially with algorithmic bias, transparency, and ethical considerations. Algorithmic bias arises when the data utilised to train AI algorithms embodies past prejudices, resulting in inequitable outcomes. If an AI model is trained on data from a corporation that has historically favoured candidates from a specific demographic, it may unintentionally perpetuate similar prejudices in its evaluations. This prompts significant enquiries on the ethical application of AI in recruitment and the necessity for transparency in the design and implementation of AI models.

This chapter aims to examine the role of AI-driven ranking models in contemporary recruitment, emphasising their advantages, limitations, and ethical considerations. As discussed above, AI-driven ranking models have substantial benefits, including enhanced efficiency, data-informed decision-making, and the capacity to mitigate bias in recruitment. Nonetheless, they also pose obstacles, especially in terms of algorithmic bias, insufficient transparency, and privacy issues.

The research questions directing this chapter are as follows:

- How does AI ranking enhance the recruitment process, and what distinct advantages do they provide?
- What are the primary problems and ethical implications related to the utilisation of AI-driven models in recruitment?
- What measures can organisations take to guarantee that AI-driven recruitment tools are equitable, transparent, and devoid of bias?

This study investigates the influence of AI-driven ranking models on recruitment, intending to enhance the discourse on the future of work and the ethical application of AI in human resource practices. The objective is to equip organisations with the knowledge and tools necessary to use AI's potential while maintaining fairness, transparency, and inclusivity in recruiting processes.

The recruitment process has evolved significantly with the integration of AI and automation, offering new efficiencies in candidate selection (Tambe *et al.*, 2019; Bersin, 2018). AI tools are increasingly being utilized to streamline candidate screening by analyzing large volumes of data, thus reducing human bias and improving decision-making accuracy (Jain &Awasar, 2020; Raghavan *et al.*, 2020). However, these advancements come with challenges. For example, Amazon's attempt to deploy an AI recruiting tool was halted when it was found to exhibit bias against female candidates (Dastin, 2018). Despite such setbacks, experts argue that AI has the potential to create a more inclusive and efficient hiring process if deployed with ethical safeguards (Binns, 2018; Marr, 2018). As Brynjolfsson and McAfee (2014) note, while automation promises to enhance productivity, it is essential for organizations to address the risks of algorithmic biases and ensure AI systems are designed and implemented responsibly. Furthermore, automation in recruitment is not without its risks, as organizations must balance efficiency with fairness (Agarwal & Dhar, 2014; Westerman *et al.*, 2014).

THE EFFICACY OF AI-DRIVEN RANKING MODELS

AI-driven ranking algorithms have transformed recruitment by automating candidate evaluation and selection processes. Central to AI-driven recruiting systems are machine learning models, natural language processing tools, and various AI methodologies aimed at analysing human language and identifying patterns in resumes, job descriptions, and other candidate data.

The Mechanisms of Artificial Intelligence in Recruitment

The operation of AI-driven ranking models can be delineated into five essential phases: data acquisition, training, model construction, and assessment (Hewage, 2023). These models utilise previous recruiting data, resumes, job descriptions, and other pertinent facts to rank candidates efficiently.

Data Acquisition

AI-driven ranking models necessitate extensive datasets for optimal performance. These datasets comprise resumes, job descriptions, performance evaluations, interview transcripts, and candidate social media profiles. Organisations frequently gather such data from applicant tracking systems, professional networking platforms (such as LinkedIn), and external recruitment services (Hosain, 2020). These statistics must be organised to enable AI models to analyse trends and discern critical indications of job performance, qualifications, and candidate suitability.

Data Preprocessing

Prior to training an AI model, raw data must undergo preprocessing to guarantee consistency and precision. Preprocessing entails eliminating extraneous information, addressing absent data, and transforming unstructured material (*e.g.*, resumes in diverse forms) into a structured manner. This phase guarantees that the data is sanitised and prepared for processing and learning by machine learning algorithms.

Training

Following data preprocessing, the subsequent stage is to train the AI models. Training entails inputting data into machine learning algorithms that discern patterns and correlations between candidate characteristics (*e.g.*, qualifications, experience) and favourable hiring results. During training, models refine their decision-making processes utilising historical hiring data. For instance, if a given organisation routinely recruits applicants with particular educational qualifications or professional experience, the AI model will learn to emphasise those characteristics when evaluating prospects.

Model Construction

Following training, the AI system develops a rating model derived from the patterns identified in the training data. This model can thereafter be utilised on new candidates to assess their appropriateness for a position. Model construction involves the selection and refinement of machine learning techniques for ranking, like decision trees, neural networks, or logistic regression models.

Assessment

Following the construction of the model, it is imperative to examine it to ascertain correctness and equity. This entails evaluating the model on novel datasets (referred to as validation or test datasets) to assess its efficacy in ranking candidates. Model evaluation also emphasises the necessity of eliminating bias in the AI system, which is essential for preventing unethical or inequitable employment practices.

Upon construction and evaluation, AI-driven ranking models can be implemented in real-time recruitment processes to assess candidate resumes, job descriptions, and additional inputs, generating ranked lists of applicants for recruiters' consideration.

Machine Learning and Data Patterns in Candidate Selection

Machine learning algorithms constitute the foundation of AI-driven ranking models in recruitment. These algorithms examine extensive datasets to identify trends and forecast a candidate's appropriateness for a position based on historical data (Garg *et al.*, 2022). Machine learning enables AI systems to automate the laborious task of resume screening, hence enhancing the efficiency and data-driven nature of the recruitment process.

Machine learning algorithms are engineered to discern patterns in resumes and job descriptions. Resumes are evaluated by algorithms that identify certain credentials, abilities, experiences, and educational backgrounds that correspond with job criteria. They examine the content and structure of job descriptions to identify essential talents and qualifications. The objective is to align candidates whose resumes exhibit comparable patterns to those of previously successful employees.

There is also Natural Learning Processing in AI-based ranking models to comprehend and evaluate human discourse. Resumes and job descriptions frequently utilise unstructured text formats, indicating a deficiency in the rigorous organisation characteristic of databases (Abedi, 2022). Natural Learning Processing enables AI systems to comprehend documents by deconstructing sentences, recognising keywords, and extracting pertinent information from the text. It can evaluate the context of specific word usage, aiding the AI model in comprehending subtleties in candidate profiles or job advertisements. NLP can differentiate between a candidate who has "worked with" technology and one who has "mastered" it (Bhukya, 2021), hence enhancing the accuracy of the model's analysis.

After the machine learning model has examined patterns within the data, it renders data-driven decisions to prioritise candidates. The decisions are founded on the model's acquired correlations between candidate characteristics and hiring efficacy. Increased data access enhances the model's decision-making capabilities, resulting in improved accuracy in candidate ranking.

Principal Algorithms in AI Recruitment

A variety of machine learning methods are frequently employed in AI-driven ranking models, each possessing distinct advantages and disadvantages. Prominent algorithms employed in recruitment encompass decision trees, neural networks, and logistic regression models.

Decision Trees

A decision tree is a widely utilised algorithm in recruitment models because of its straightforwardness and clarity of interpretation. In a decision tree, the AI system renders judgments by partitioning the data into branches according to particular attributes (Choudhary *et al.*, 2023). A decision tree may initially categorise individuals according to education, followed by experience, and subsequently talents. Each "branch" signifies a distinct decision pathway, ultimately culminating in a rating of candidates. The primary benefit of decision trees is their interpretability, enabling recruiters to comprehend the rationale behind a candidate's higher ranking. Nonetheless, decision trees can occasionally exhibit overfitting, resulting in commendable performance on training data but subpar results on novel, unknown data.

Neural Networks

Neural networks are sophisticated algorithms adept at analysing intricate data. Neural networks have layers of interconnected nodes (neurones) that analyse information and discern patterns within data. These networks are especially proficient at managing extensive datasets and can discern intricate links among variables. Neural networks in recruiting can evaluate many inputs (*e.g.*, resumes and interview transcripts) and generate precise predictions regarding candidate suitability (Hemamou *et al.*, 2021). Neural networks are frequently termed "black boxes" due to their interpretative complexity (Dobson, 2023). Recruiters may lack comprehension regarding the rationale behind a neural network's preferential ranking of one candidate over another, hence eliciting worries about transparency.

Logistic regression is a prevalent technique in AI recruitment due to its simplicity and efficacy in binary classification tasks, such as determining a candidate's suitability. It functions by assessing the likelihood of a candidate's success in a specific position based on the provided data. Logistic regression models are readily interpretable, rendering them favoured in recruitment contexts where transparency and elucidation are paramount (Jacobs *et al.*, 2021). Nonetheless, logistic regression may exhibit suboptimal performance with extremely intricate datasets, necessitating the utilisation of more sophisticated algorithms such as neural networks.

Artificial Intelligence Instruments for Curriculum Vitae Evaluation

AI-driven ranking models have transformed resume screening, enabling recruiters to quickly manage extensive quantities of resumes and effortlessly find leading applicants. These models utilise a synthesis of keyword extraction, natural language processing, and machine learning to assess resumes and align them with job descriptions.

Keyword Extraction

AI models initiate the process by analysing resumes for pertinent keywords that correspond to the job description. The keywords encompass certain talents, credentials, certifications, and job titles pertinent to the function. If a job description specifies "Python programming" as a requisite skill, the AI model will prioritise resumes containing this term (Kabir, 2024).

Candidate Ranking

According to Qualifications and Experience: Following keyword extraction, AI models evaluate and rank candidates based on their qualifications and experience. The models evaluate the candidate's professional trajectory, academic qualifications, and competencies to assess their alignment with the job's specifications. A candidate's ranking will increase in accordance with the degree of alignment between their profile and the job description.

Natural Learning Process

facilitates contextual analysis, enabling AI applications to go beyond mere keyword matching by examining the context around the terms. An applicant might claim "Python programming" as a competence, although NLP can evaluate if the individual possesses substantial expertise with Python or has only referenced it superficially. This contextual analysis enhances the AI model's ability to deliver more precise rankings.

Difficulties in Analysing Intricate Human Inputs

Notwithstanding the progress in AI-driven rating algorithms, these systems continue to encounter difficulties in deciphering intricate and nuanced human inputs, like resumes including unorthodox career trajectories or creatively formatted structures.

Candidates with non-linear career paths including those who have transitioned between industries or experienced job interruptions may not be appropriately assessed by AI models. These algorithms frequently favour candidates with conventional professional trajectories, potentially neglecting those who offer useful and unique experiences.

Certain individuals employ unconventional resume styles to highlight their originality or distinctive abilities. AI-driven rating models may encounter difficulties in interpreting resumes that deviate from the expected conventional structure (Rigotti & Fosch-Villaronga, 2024). A graphic designer might present a visually innovative résumé that presents essential information in atypical locations, complicating the extraction and analysis of pertinent data by AI algorithms.

The efficacy of AI models is contingent upon the quality of the data utilised for training. If the training data embodies prejudices from prior employment practices, such as preferential treatment towards specific demographics or educational qualifications, the AI model may perpetuate these biases. This poses a substantial ethical dilemma, as biased AI models may lead to inequitable employment outcomes.

ADVANTAGES OF AI-ENHANCED RANKING MODELS

AI-driven ranking algorithms are transforming the recruitment environment by automating labour-intensive operations, improving data-informed decision-making, alleviating administrative loads on recruiters, and fostering diversity in hiring. As organisations further integrate AI into recruiting, the advantages of these systems manifest in enhanced efficiency, superior quality of hires, and a more equitable, objective hiring process. This section examines the several benefits of employing AI-driven ranking algorithms in recruitment, substantiated by examples, case studies, and data.

Enhanced Recruitment Efficiency

The enhancement of recruitment efficiency is one of the most notable advantages of AI-driven ranking models. Conventional recruitment techniques, which entail the manual assessment of resumes, holding interviews, and sifting through prospects, can be arduous and time-consuming. Artificial intelligence streamlines this process, allowing organisations to expedite the recruitment of candidates with reduced effort.

Accelerated Resume Evaluation

AI systems can assess hundreds of resumes in a significantly shorter duration than a human recruiter. These models are designed to swiftly recognise keywords, abilities, and qualifications pertinent to the job description, thereby refining extensive candidate pools into a concise shortlist. Deloitte's analysis indicates that organisations employing AI in their recruitment processes saw a 60-70% decrease in the duration required to evaluate candidates (Hassan, 2022).

Unilever, a multinational consumer products corporation, employed AI-driven tools to optimise its graduate recruitment process. Utilising AI for resume evaluation and video interviews, Unilever achieved a 75% reduction in recruitment time (Lisa & Talla Simo, 2021). The AI technology assessed candidates' facial expressions, vocal tone, and diction during video interviews, enabling the organisation to evaluate thousands of applicants more effectively. Consequently, Unilever diminished interview duration by 30,000 hours per year and employed graduates with superior performance metrics (Agnihotri & Bhattacharya, 2024).

Predictive Analytics for Candidate Matching

AI-based rating algorithms employ predictive analytics to evaluate a candidate's prospective performance in a certain position. This capacity allows organisations to identify high-potential individuals early in the recruitment process, hence minimising the time allocated to superfluous interviews or assessments. AI algorithms can forecast a candidate's suitability for an organisation by analysing their historical performance, career progression, and cultural compatibility.

IBM employs AI-driven tools to enhance its recruitment process. IBM's AI system employs predictive analytics to assess candidates according to their probability of success within the organisation (Allal-Chérif *et al*., 2021). The AI algorithm aligns candidates' credentials with job descriptions and forecasts their likelihood for sustained success, markedly decreasing the recruiting cycle and enhancing employee retention rates.

Improved Data-Driven Decision Making

One benefit of AI-driven ranking algorithms is their capacity to improve data-informed decision-making in recruitment. AI systems can analyse extensive datasets, discern patterns, and offer insights that facilitate better-informed hiring decisions for recruiters. Organisations may enhance the quality of their hiring, decrease turnover rates, and refine their recruitment strategy by utilising data analytics.

Enhanced Quality of Hires

AI algorithms evaluate historical data from prior successful hires to discern the characteristics and qualifications most conducive to effective hiring outcomes. These insights empower recruiters to make data-informed decisions in candidate rating, resulting in superior hiring quality. A 2020 study revealed that organisations employing AI tools in recruitment observed a 50% enhancement in the calibre of their personnel (Suneetha *et al.*, 2024).

Google: Google has consistently been at the forefront of utilising data analytics for recruitment purposes. The business created an AI-powered tool called "Google Hire," designed to assist recruiters in evaluating individuals' historical performance, employment background, and cultural compatibility (Khan, 2017). Google Hire uses machine learning to evaluate candidates according to Key Performance Indicators (KPIs) derived from successful employees. Consequently, Google has markedly enhanced the precision of its recruitment choices and diminished attrition rates among recent employees.

Candidate Data Analysis

AI systems aggregate and evaluate diverse candidate data, encompassing social media profiles, professional experience, educational background, and online portfolios. Through data analysis, AI may furnish recruiters with enhanced insights into candidates' qualifications and prospective work performance. These insights assist recruiters in making better-informed selections and selecting individuals who are more likely to excel in their positions.

Reports and Statistics

A report by McKinsey *et al.* indicates that organisations implementing AI in recruiting see a 35% decrease in employee attrition, attributed to improved candidate matching and data-driven hiring decisions (Veglianti *et al.*, 2023) Moreover, AI tools assist organisations in identifying skill deficiencies within their workforce and formulating focused recruitment plans to rectify these deficiencies.

Alleviation of Administrative Load on Recruiters

AI-driven rating models substantially alleviate the administrative workload for recruiters, enabling them to concentrate on more strategic endeavours such as cultivating relationships with premier talent, strengthening corporate branding, and improving applicant experience.

A principal advantage of AI in recruiting is its capacity to automate redundant processes, including resume screening, candidate matching, and interview scheduling. AI-driven solutions may now manage duties that formerly occupied a substantial portion of recruiters' time, enabling them to concentrate on more value-added operations.

Hilton Hotels utilised AI-driven recruitment technologies to streamline its employment process. The AI system evaluates resumes, does preliminary applicant assessments, and arranges interviews. This technology diminished the administrative burden on Hilton's recruiters by 75%, allowing them to allocate more time to interacting with premier applicants and enhancing the entire candidate experience (Khan *et al.*, 2023).

With AI managing administrative duties, recruiters can concentrate on strategic activities, including cultivating talent pipelines, engaging passive prospects, and fostering relationships with premier talent. AI solutions furnish recruiters with critical data insights, allowing them to discern trends in the talent market and enhance their recruitment efforts accordingly.

AI tools can assist recruiters in tailoring their interactions with candidates. AI systems can provide tailored communications and engagement methods by examining candidates' interests, career aspirations, and previous experiences with the organisation. This customisation fosters more robust interactions with prospects and enhances the probability of acquiring premier talent.

The Role of AI in Enhancing Workforce Diversity

AI-driven ranking algorithms are essential for enhancing diversity in recruiting by mitigating human biases and facilitating a more equitable, objective hiring process. Conventional recruitment techniques frequently exhibit unconscious biases, leading to insufficient workforce diversity. Conversely, AI systems utilise anonymised data and objective metrics to assess candidates, aiding organisations in fostering diversity and inclusion.

Numerous AI-based rating methods are engineered to anonymise candidate information during the screening procedure. This indicates that personal data, including names, gender, and ethnicity, is eliminated from the preliminary phases of candidate assessment. AI solutions facilitate the anonymisation of candidate data, ensuring that assessments are based exclusively on qualifications and experience, hence mitigating potential bias from personal attributes.

Blendoor is an artificial intelligence-driven recruitment tool aimed at mitigating prejudice in the hiring process. The platform conceals candidates' personal information and emphasises their qualifications, talents, and experience. Organisations utilising Blendoor have shown considerable enhancements in workforce diversity, as AI technology guarantees equitable and objective evaluation of all candidates (Chen, 2023). Blendoor has excelled in advancing gender and racial diversity inside technology firms, an industry often characterised by its deficiency in diversity.

AI tools can facilitate blind recruiting, a practice in which recruiters lack access to candidates' personal information. Blind hiring emphasises candidates' abilities, experience, and job-related credentials, hence removing the possibility of biases associated with age, gender, ethnicity, or socioeconomic status. Organisations that implement blind recruiting processes frequently observe more diversity within their candidate pools and a more inclusive working culture.

Numerous studies indicate that AI-powered recruitment tools can mitigate gender prejudice in the employment process. A World Economic Forum study revealed that AI tools aimed at mitigating gender prejudice in recruitment led to a 20% rise in the employment of women for technical positions (Collett *et al.*, 2022). By emphasising candidates' qualifications and removing gender-related biases, AI techniques can assist organisations in fostering gender diversity and establishing more inclusive workplaces.

Although AI can enhance diversity, it is crucial for organisations to guarantee that the algorithms employed in recruitment are devoid of prejudice. AI systems are contingent upon the quality of their training data; if this data embodies past biases, the AI model may replicate those biases. Consequently, it is imperative for organisations to consistently evaluate their AI systems to guarantee the promotion of equitable and ethical employment practices.

CHALLENGES AND ETHICAL ISSUES IN AI-DRIVEN RANKING MODELS

Although AI-driven ranking models have enhanced efficiency, automation, and data-informed decision-making in recruitment, they present hurdles and ethical dilemmas. These technologies may result in adverse consequences, including algorithmic bias, insufficient transparency, and privacy concerns. The implementation of AI in recruitment has ignited extensive ethical discussions over the dehumanisation of the hiring process and the substitution of human judgment with algorithms. This section examines these difficulties comprehensively.

Algorithmic Bias and Equity Concerns

Algorithmic bias arises when AI systems, trained on historical or prejudiced data, sustain discriminating results. In recruitment, this may lead to specific groups, such as women or minority candidates, being unjustly disadvantaged.

The efficacy of AI systems is contingent upon the quality of their training data. If the training data embodies historical biases, such as gender or racial discrimination, the AI model may unintentionally reproduce these biases in employment decisions. For instance, if an AI system is trained on resumes from a mostly male IT workforce, it may develop a bias favouring male candidates over female prospects, despite equivalent qualifications. This may sustain prevailing disparities in the labour market.

A notable instance of algorithmic bias in AI-driven recruiting is Amazon's AI tool, intended to automate the hiring process by evaluating prospects. Nonetheless, the technology was discovered to systematically bias against female applicants. The AI was trained on resumes provided to Amazon over a decade, predominantly from male applicants. The AI consequently exhibited a preference for male candidates and discriminated against resumes containing the term "women," such as "women's chess club captain." Amazon ultimately discontinued the product, underscoring the risks associated with biased training data (Andrews & Bucher, 2022).

AI techniques albeit lacking apparent bias may still result in inadvertent discrimination. Certain demographic groupings may be inadequately represented in the training data, resulting in the AI making less accurate predictions regarding candidates from these categories. This may lead to the exclusion of suitable applicants only due to their failure to conform to the patterns recognised by the AI's training.

Confronting algorithmic bias necessitates meticulous consideration of the quality and diversity of the training data employed in the development of AI models. Organisations must guarantee that their AI systems are trained on data that accurately represents all demographic groups and that the models undergo frequent audits for equity. Debiasing algorithms and fairness constraints can be employed to alleviate bias in AI-driven recruitment models.

The "Black Box" Dilemma: Absence of Transparency in AI Decision-Making

A primary worry of AI-driven ranking methods is their opacity. This issue, commonly known as the "black box" dilemma, occurs when AI systems render decisions without offering transparent justifications for the rationale behind those decisions (Von Eschenbach, 2021). The absence of transparency in recruitment can significantly impact accountability and equity.

Numerous AI models, especially those utilising machine learning, function in ways that are not readily comprehensible to humans. These models employ intricate algorithms to examine extensive datasets and discern trends; however, the methodology behind their decision-making remains obscured to users. Consequently, recruiters may be unaware of the reasons behind the varying rankings of prospects or the rationale for selecting one candidate over another.

The absence of transparency in AI-driven conclusions complicates the accountability of recruiters and organisations about biased or inequitable recruiting practices. When an applicant is declined due to an AI-generated ranking, it may be infeasible for the recruiter to elucidate the rationale behind the choice. The absence of accountability is especially concerning in instances where AI has rendered biased decisions, as there is no definitive method to pinpoint or rectify the origin of the bias.

In response to the black box dilemma, researchers and developers are focusing on explainable AI (XAI), which seeks to enhance the transparency and interpretability of AI systems (Gerlings *et al.*, 2020). XAI aims to furnish transparent elucidations regarding the decision-making processes of AI models, enabling users to comprehend and have confidence in the results. In recruitment, XAI could facilitate the assurance that AI-driven rating algorithms render equitable and reasonable decisions.

Mitigating Data Privacy and Security Issues

AI-driven ranking models depend on the aggregation and examination of extensive personal data, encompassing resumes, social media profiles, and video interviews. This creates substantial concerns regarding data privacy and security, as candidates' personal information is amassed, retained, and scrutinised by AI systems (Peña *et al.*, 2023).

A primary problem in AI-driven recruiting is the acquisition of candidates' personal data with their consent and its utilisation in accordance with privacy legislation. In numerous instances, candidates may lack complete awareness regarding the utilisation of their data by AI systems or the specifics of the information being gathered. The absence of transparency may raise apprehensions regarding data exploitation and breaches of privacy.

Organisations employing AI-driven recruitment technologies must ensure compliance with data privacy legislation, including the General Data Protection Regulation in Europe (Aloisi, A. 2024). The General Data Protection Regulation mandates that enterprises secure explicit consent from individuals before the collection of personal data and furnish clear explanations regarding the intended use of such data. Noncompliance with GDPR may lead to substantial penalties and harm to a company's reputation. In India, the Personal Digital Data Protection Act, 2023, also has similar provisions to protect privacy and ensure the safety and security of personal data through regulatory compliance (Harinath, G. 2023).

The implementation of AI in recruitment presents issues regarding data protection. AI-driven tools frequently retain and manage substantial quantities of sensitive personal data, rendering them susceptible to cyberattacks. Data breaches may lead to the disclosure of candidates' personal information, encompassing their contact data, work history, and pay information. Organisations must allocate resources towards comprehensive cybersecurity protocols to safeguard this information and avert unauthorised access.

To mitigate privacy concerns, several AI-driven recruitment models include data anonymisation techniques that eliminate Personally Identifying Information (PII) from the data prior to processing. This guarantees the protection of candidates' privacy while permitting the AI to examine the pertinent data. Nonetheless, anonymisation is not infallible, and there is a possibility that anonymised material may be re-identified.

Beyond adherence to privacy legislation, organisations must also contemplate the ethical ramifications of employing candidates' personal data for recruitment objectives. Candidates may feel uneasy about the prospect of their social media engagement, facial expressions in video interviews, or other personal data being scrutinised by AI. Organisations must find an equilibrium between utilising AI's potential and safeguarding candidates' privacy rights.

Ethical Quandaries in the Application of AI for Recruitment

The application of AI in recruiting presents significant ethical concerns around the influence of technology on the hiring process and the risk of dehumanising recruitment practices.

A primary ethical worry regarding AI-driven recruitment is the possible dehumanisation of the process. Recruitment has historically been a predominantly human-centric process, wherein recruiters cultivate connections with prospects, evaluate their interpersonal abilities, and render subjective assessments of their suitability for an organisation. In contrast, AI depends on objective facts and algorithms for decision-making, potentially reducing candidates to mere numerical values and buzzwords. This may result in a more impersonal and transactional recruitment procedure.

HireVue is an AI-based video interview platform that employs facial recognition and audio analysis to evaluate candidates' appropriateness for a position (Ajunwa, 2021). Although the platform asserts that it offers impartial evaluations of candidates' communication abilities and emotional intelligence, some contend that it simplifies candidates to just facial expressions and vocal tone, neglecting the intricacies of human connection. Furthermore, candidates may experience discomfort regarding the prospect of being evaluated by an algorithm instead of a human recruiter.

AI-driven recruitment tools elicit apprehensions around the substitution of human judgment with machine decision-making. Although AI can process extensive data and recognise patterns, it may lack the capacity to evaluate the intangible attributes that determine a candidate's suitability for a position, such as emotional intelligence, creativity, or cultural alignment. Excessive dependence on AI in recruitment may result in hiring judgments that favour technical talents at the expense of softer, more subjective attributes.

A further ethical difficulty pertains to the issue of accountability in AI-driven recruitment. Who bears responsibility for a judgment made by an AI-driven ranking mechanism that results in a candidate's rejection? Is it the recruiter utilising the tool, the corporation that created the AI system, or the AI itself? The absence of explicit accountability in AI-driven recruitment engenders apprehensions over fairness and the possibility of candidates being unjustly dismissed without recourse.

To mitigate these ethical issues, numerous experts endorse a hybrid recruitment model, wherein AI aids human recruiters instead of supplanting them. Artificial intelligence can perform functions such as resume evaluation and candidate prioritisation; however, the ultimate hiring decision must be rendered by a human recruiter who considers both the AI's suggestions and their personal discernment. This method guarantees a comprehensive evaluation of candidates, incorporating both quantitative data and subjective assessment in the decision-making process.

Approaches for Ethical Artificial Intelligence Development in Recruitment

Creating AI models that are equitable and transparent is essential for mitigating ethical issues in recruitment. Equity guarantees that AI systems do not reinforce prejudices, whereas openness promotes accountability and trustworthiness. Attaining these objectives necessitates the implementation of ethical frameworks, the development of explainable AI, and the ongoing evaluation of AI models.

A highly effective method for ensuring the ethical development of AI systems is to comply with predefined frameworks and norms. The European Union's Ethics Guidelines for Trustworthy AI delineate a thorough framework of principles designed to guarantee that AI systems are legal, ethical, and resilient (Hickman & Petrin, 2021). These standards underscore concepts including respect for human autonomy, harm prevention, equity, and clarity. Integrating these principles into AI development can assist organisations in averting unethical results in recruitment.

The EU rules that AI systems be constructed under human supervision, guaranteeing that humans maintain authority over the ultimate decision-making process (de Almeida *et al.*, 2021). The rules promote variety in training data and recommend regular audits to verify that AI systems adhere to ethical principles.

AI models must be subjected to consistent auditing and testing to detect potential biases or unexpected consequences. Audits must concentrate on assessing the alignment of the AI system with ethical standards and determining if the system yields equitable and impartial outcomes across all demographic groups. Moreover, ongoing testing with empirical data might assist developers in recognising and rectifying ethical concerns prior to their escalation.

Approaches, including fairness requirements and bias correction algorithms, can assist in rectifying biases inside AI systems. Fairness restrictions can be integrated into the model's architecture, guaranteeing that the AI does not preferentially advantage one group over another in its decision-making processes. These limits can be harmonised with legal mandates, such as equal employment opportunity standards, assuring adherence to anti-discrimination statutes.

The precision and equity of AI models in recruitment mostly rely on the quality and variety of the training data utilised. AI algorithms based on biased or unrepresentative data may unintentionally reproduce and exacerbate historical injustices. Organisations must prioritise data diversity and establish rigorous data audits and validation methods to mitigate these threats.

A crucial step in developing equitable AI models is to guarantee that the training data encompasses a broad spectrum of demographic groups. Datasets that disproportionately reflect specific demographics, such as men in technology sectors, might result in biased AI systems that unjustly harm underrepresented individuals, including women and people of colour. To prevent this, organisations must gather varied data that represents the wide range of applicants across gender, ethnicity, age, and other attributes.

A worldwide technology firm identified that its AI recruitment system was systematically rejecting female candidates for software engineering positions (Pisanelli, 2023). Upon examining the training data, the organisation discovered that the dataset utilised for AI training predominantly consisted of resumes from male candidates. By augmenting the dataset to encompass a more heterogeneous collection of resumes, the company successfully mitigated bias and enhanced the equity of the AI system.

Organisations must establish stringent data auditing and validation protocols to guarantee that AI systems are trained on equitable and impartial data. Data audits entail examining the training data to detect and rectify any biases or disparities. This may involve examining the representation of various demographic groups and verifying that the data appropriately represents the qualifications and abilities required for the position.

When real-world data is scarce or biased, organisations may resort to synthetic data as a remedy. Synthetic data is intentionally created data that simulates real-world data while being engineered to be more representative and impartial. Organisations may enhance the variety and equity of their AI systems and mitigate the risk of data scarcity by augmenting training datasets with synthetic data.

Organisations must maintain adherence to data privacy regulations, such as the Digital Personal Data Protection Act 2023, while enhancing data diversity. The utilisation of personal data in AI training necessitates specific consent, and organisations must adopt methods to anonymise or pseudonymize candidate data to safeguard human privacy. Consistent evaluation of data protection methods helps avert ethical violations and uphold candidates' trust.

AI should not be perceived as a substitute for human judgment in recruitment, but rather as an instrument to aid HR professionals in making more informed judgments. Human supervision is crucial for the ethical and effective utilisation of AI-driven recruitment technologies.

AI can improve recruitment by automating monotonous processes like resume screening; nevertheless, it should not supplant human intuition, empathy, and judgment. Human Resources experts introduce a personal dimension to the recruitment process, essential for evaluating a candidate's cultural compatibility, interpersonal skills, and growth potential. By employing AI as a means of enhancement rather than replacement, HR managers can capitalise on the advantages of both human discernment and data-informed insights.

A primary responsibility of HR professionals is to supervise the judgments rendered by AI systems, ensuring their alignment with ethical standards and organisational values. This entails the systematic evaluation of AI-generated results, recognising any biases, and acting as needed to avert inequitable conclusions. Human Resources specialists must also furnish feedback to AI developers to enhance the precision and equity of the models.

Human Resources professionals must exercise caution regarding excessive dependence on AI-driven recruitment technologies. Although AI can discern trends in applicant data, it may not fully encapsulate the intricacies of a candidate's experience or potential. Human Resources personnel ought to regard AI recommendations as a component of the decision-making process, rather than the exclusive factor influencing recruiting results. Organisations can guarantee the responsible and ethical use of AI tools by upholding human oversight.

HR personnel must be educated about the functionality of AI technologies, encompassing their limitations and potential biases, to successfully supervise these systems. Equipping HR teams with training on AI ethics, data protection, and bias mitigation would enable them to make educated judgments regarding the use of AI in recruitment. By cultivating a culture of perpetual learning, organisations may guarantee that their HR personnel remain attentive to maintaining ethical standards in AI-facilitated recruitment.

For instance, organisations such as Google have implemented AI ethics training programs for HR workers, enabling them to comprehend the ethical implications of utilising AI in recruitment processes (Kshetri, 2021). These programs underscore the significance of transparency, equity, and data security, ensuring that HR professionals are adequately prepared to manage AI systems proficiently.

FUTURE PERSPECTIVE: THE SIGNIFICANCE OF COLLABORATION IN AI-ENHANCED RECRUITMENT

As AI transforms recruitment, its future efficacy hinges on a collaborative methodology. The increasing intricacy of AI systems along with the ethical and legal dilemmas they pose necessitates a collaborative effort between technical and non-technical parties. By promoting collaboration among data scientists, HR professionals, legal experts, and ethicists, organisations can guarantee that AI-driven recruitment is both innovative and responsible. This part will examine the significance of collaboration, address the necessity for industry standards and best practices, and offer a prospective analysis of how future technologies will continue to transform AI-driven recruitment.

Collaborative Interdisciplinary Approaches for Ethical Artificial Intelligence

Ethical AI development in recruitment necessitates collaboration beyond technologists alone. Although data scientists and software engineers develop AI systems, the ethical, legal, and social ramifications of these technologies surpass the technological domain. Organisations must foster collaboration between technical teams and non-technical stakeholders, such as HR professionals, legal experts, and ethicists, to develop AI solutions that are both effective and ethical. Each group contributes distinct ideas and skills, essential for tackling the complex difficulties presented by AI-driven recruitment.

Data scientists and software engineers are pivotal in the creation and development of AI-driven recruitment solutions. They are tasked with developing machine learning algorithms and Natural Language Processing (NLP) systems that drive AI technologies, in addition to training these models on extensive datasets. Nevertheless, technical teams frequently prioritise the efficiency and accuracy of AI systems, while placing little emphasis on the ethical and societal ramifications of their endeavours. Collaboration with non-technical stakeholders is crucial at this juncture.

HR experts serve as ethical gatekeepers, with a profound comprehension of recruiting procedures, candidate experience, and organisational culture, which enables them to supervise the ethical application of AI in recruitment. By collaborating with data scientists, HR professionals may guarantee that AI systems are developed in accordance with the organization's values and ethical principles. HR experts can evaluate AI-generated recruiting recommendations, ensuring that human judgment is integral to the recruitment process. Furthermore, HR teams can assist in identifying potential biases in AI models and collaborate with data scientists to address these concerns.

Legal professionals are essential in ensuring that AI-driven recruitment tools adhere to employment rules, data protection standards, and anti-discrimination policies. In areas subject to the General Data Protection Regulation or the Digital Personal Protection Data, 2023, legal practitioners can counsel on the collection and processing of candidate data while upholding privacy rights. Legal specialists assist organisations in navigating the intricate framework of anti-discrimination legislation, ensuring that AI systems do not unintentionally contravene equal employment opportunity regulations. Through collaboration with data scientists and human resources professionals, legal experts can guarantee that AI-driven recruitment tools adhere to compliance standards and promote fairness.

Ethicists offer significant perspectives on the overarching ethical ramifications of employing AI in recruitment. They can assist organisations in addressing ethical challenges, such as reconciling the necessity for efficiency with the risk of dehumanisation in recruitment. Ethicists can assist in establishing frameworks to assess the fairness and transparency of AI systems, ensuring these technologies foster social good instead of reinforcing inequality. Ethicists in interdisciplinary collaboration can counterbalance exclusively technical or operational viewpoints, promoting a more comprehensive approach to AI development.

Organisations can facilitate interdisciplinary collaboration by forming formal working groups or committees that unite data scientists, HR specialists, legal experts, and ethicists. These organisations may convene periodically to evaluate AI systems, address ethical and legal issues, and recommend measures to enhance fairness, transparency, and accountability. Incorporating ethical issues into the development process from the beginning enables organisations to ensure that their AI-driven recruitment solutions are innovative and responsible.

Industry Standards and Optimal Practices

As artificial intelligence transforms recruitment, there is an imperative for comprehensive industry standards and best practices to guarantee the responsible use of AI tools. Now, there is an absence of formalised regulations regarding the utilisation of AI in recruitment, resulting in varied practices and a deficiency in accountability. Through the establishment of industry standards, organisations may delineate explicit expectations for ethical AI utilisation, ensuring that all stakeholders from data scientists to HR professionals are accountable to uniform ethical and legal standards.

Although widely agreed standards for AI in recruiting are lacking, some organisations and regulatory entities have initiated the formulation of recommendations for the ethical application of AI. The European Union's Ethics Guidelines for Trustworthy AI delineate principles designed to guarantee that AI systems are legal, ethical, and resilient (Díaz-Rodríguez *et al.*, 2023). These principles underscore the significance of openness, equity, and accountability, providing a helpful resource for organisations aiming to adopt ethical AI tools in recruitment.

For instance, the IEEE has established a framework for ethical AI development referred to as Ethically Aligned Design. These principles underscore the necessity for human-centric AI systems that prioritise equity, accountability, and transparency. By implementing these criteria, organisations may guarantee that their AI-driven recruitment solutions are developed with ethical considerations.

The Prospects of AI-Enhanced Recruitment

The future of AI-driven recruitment presents promising prospects, as emerging technologies are poised to further revolutionise organisational talent acquisition. As AI systems mature, the emergence of innovative solutions to improve the recruitment process is anticipated, including AI-driven interviews, psychometric evaluations, and automated reference verification (Kazim *et al.*, 2021). Nonetheless, these advancements also introduce new concerns, especially regarding equity, transparency, and human supervision.

A notable trend in AI-driven recruitment is the implementation of AI-enhanced interviews. These instruments evaluate candidates' facial expressions, vocal tone, and diction during interviews to gauge attributes such as emotional intelligence, problem-solving skills, and cultural compatibility. AI-driven interviews can enhance the recruiting process and offer impartial evaluations of candidates; nevertheless, they also provoke questions about prejudice and equity. AI systems may misread cultural variances in communication styles or unjustly disfavour candidates who exhibit nervousness or discomfort during virtual interviews.

Organisations must guarantee that AI-powered interviews are developed to be equitable and transparent to ensure ethical usage. This may involve furnishing candidates with explicit elucidations of the AI system's functionality, in addition to presenting alternative interview styles for those who are apprehensive about AI evaluations.

Artificial intelligence is utilised to administer psychometric evaluations that test candidates' personality characteristics, cognitive capabilities, and emotional intelligence. These evaluations can yield significant insights into individuals' potential performance in particular positions, enabling organisations to make more educated recruiting choices. Nonetheless, similar to AI-driven interviews, psychometric evaluations provide ethical dilemmas, especially regarding prejudice and the dehumanisation of the hiring process.

To mitigate bias in psychometric evaluations, organisations must guarantee that their AI systems are trained on varied datasets that encompass a broad spectrum of personalities, backgrounds, and experiences. Furthermore, organisations ought to furnish candidates with feedback regarding their evaluations, enabling them to comprehend the criteria used for evaluating their work.

An emerging trend in AI-driven recruitment is the automation of reference verification. AI systems can evaluate data from former employers, coworkers, and professional networks to appraise candidates' job history, performance, and dependability. Automated reference verification can conserve time and yield more objective insights compared to conventional reference checks. Nonetheless, it also elicits apprehensions regarding privacy and data security, along with the possibility of data mistakes.

Organisations must secure informed consent from candidates prior to executing automated reference checks to safeguard their privacy. Furthermore, organisations must establish stringent data protection protocols to avert unauthorised access to candidate information.

REAL-WORLD FAILURES IN AI HIRING: CASE STUDIES AND LESSONS LEARNED

Amazon's AI Recruiting Tool: A Lesson in Bias Amplification

In 2018, Amazon discontinued an internal AI recruiting tool after discovering it systematically discriminated against female candidates. The algorithm was trained on historical resume data submitted to Amazon, which largely came from male applicants in tech roles. As a result, the AI penalized resumes that included the word "women" (*e.g.*, "women's chess club captain") and favored male-associated language patterns (Dastin, 2018).

Lesson Learned

- **Biased historical data can replicate and even intensify discrimination** if not corrected before model deployment.
- Ensuring **gender-diverse and demographically balanced training data** is critical for fair outcomes.

HireVue and the Black Box Challenge

HireVue, a platform using AI to analyze facial expressions and vocal tone during video interviews, has faced criticism for reducing candidates to "data points" and relying on subjective traits like tone and facial movement. Critics argue this method risks **misjudging qualified candidates** based on cultural or neurodiverse expression styles (Ajunwa, 2021).

Lesson Learned

- Avoid over-reliance on **opaque AI metrics** in screening.
- Organizations must adopt **explainable AI (XAI)** methods to ensure transparency and trust.

Unintended Discrimination Through Skewed Training Data

A global tech firm discovered its AI tool was rejecting female applicants disproportionately for software roles. A post-mortem revealed the training dataset included **90% resumes from male candidates**, leading to learned gender bias. After diversifying the dataset and revalidating the model, hiring equity improved significantly (Pisanelli, 2023).

Lesson Learned

- Implement **ongoing audits** and **bias detection protocols**.
- Use **synthetic data augmentation** when real-world data lacks diversity.

CONCLUSION

In the swiftly changing realm of recruitment, artificial intelligence has surfaced as a revolutionary influence, providing unparalleled efficiency, data-driven decision-making, and the capacity to mitigate human biases in hiring. A detailed investigation has revealed several crucial points. AI-driven recruiting models offer various benefits to organisations, including the optimisation of administrative duties and the enhancement of data-driven, objective decision-making processes. These technologies enable firms to analyse extensive candidate data with exceptional speed and precision, significantly diminishing the time and expenses linked to conventional hiring practices. Organisations such as Unilever and IBM have illustrated that AI can substantially improve the recruitment process by elevating the quality of hiring, liberating HR personnel from monotonous activities, and enabling them to concentrate on strategic, human-centric endeavours. Furthermore, AI possesses the capacity to enhance diversity by eliminating human biases that frequently affect employment decisions due to subjective evaluations or unconscious inclinations.

Nonetheless, notwithstanding these advantages, AI-driven recruitment models pose considerable problems and ethical dilemmas that require resolution to guarantee their responsible and equitable application. Algorithmic bias constitutes a principal challenge. AI systems' efficacy is contingent upon the quality of their training data; biased training data can result in biased outcomes, hence perpetuating inequities and systematic discrimination in employment processes. This problem was exemplified in the notorious instance of Amazon's AI tool, which inadvertently discriminated against female applicants because it utilised historical recruiting data that mirrored the company's male-centric staff. Moreover, the opaque nature of some AI models engenders apprehensions over transparency and accountability. When AI systems render decisions without transparent explanations, users struggle to comprehend, trust, or contest these decisions, resulting in less accountability within the recruitment process.

Data privacy and security are key issues in the utilisation of AI-driven recruitment solutions. AI systems depend on extensive personal data, such as resumes, social media profiles, and psychometric evaluations, to inform employment decisions. This prompts enquiries on the methods by which companies gather, retain, and utilise candidate data, particularly in consideration of regulations such as the General Data Protection Regulation. Safeguarding candidate privacy and guaranteeing data security are crucial for fostering trust in AI systems and preventing the misuse of personal information.

In light of these issues, the creation of equitable and transparent AI models is essential. Organisations must prioritise the development of AI systems that are transparent, responsible, and devoid of prejudice. Collaboration between technical and non-technical parties is necessary to accomplish this. Data scientists and software engineers must collaborate closely with HR professionals, legal experts, and ethicists to ensure that AI tools are developed and implemented in accordance with ethical norms and legal rules. This interdisciplinary partnership can assist organisations in reconciling the technological efficiency of AI with the requirements for fairness, openness, and human oversight in recruiting.

Subsequent investigations in AI-enhanced recruitment ought to concentrate on numerous crucial domains. Addressing algorithmic prejudice must be prioritised. This can be accomplished by enhancing the quality and variety of training datasets, ensuring that AI models are trained on data that represent a broad spectrum of applicant backgrounds, experiences, and qualifications. Furthermore, organisations ought to conduct periodic audits of their AI systems to detect and rectify any biases that may develop over time. Fairness-aware machine learning techniques, designed to address bias and enhance equity, should be further investigated and incorporated into AI recruitment tools.

Secondly, promoting inclusive employment practices with AI is a vital domain for future investigation. Although AI possesses the capability to mitigate human biases, it may also unintentionally perpetuate existing disparities if not meticulously regulated. Researchers ought to concentrate on creating AI models that not only eliminate bias but also actively foster diversity and inclusion. AI technologies can be engineered to anonymise candidate data by eliminating details such as gender, age, or race that may result in biased choices. AI-driven recruiting models can assist organisations in constructing more diverse and inclusive workforces by emphasising objective factors such as skills, certifications, and experience.

Furthermore, the significance of human oversight in AI-facilitated recruitment must be underscored. Although AI tools can markedly improve the efficiency and precision of the recruitment process, they must not supplant human judgment. Human Resources professionals must maintain the capacity to evaluate and contest AI-generated recommendations, guaranteeing that ultimate hiring decisions are made with a comprehensive understanding of the candidate. Human oversight is essential for resolving the ethical concerns linked to AI, including the dehumanisation of the recruitment process and the potential reduction of candidates to mere data points. Organisations may develop recruitment procedures that are both efficient and human-centred by integrating the strengths of AI with human empathy, intuition, and experience.

There is a necessity for explicit industry standards and best practices to govern the ethical application of AI in recruitment within regulatory frameworks. The absence of formalised norms has resulted in varied procedures and diminished responsibility in AI-driven recruiting. Governments and industry organisations must collaborate to formulate extensive rules that tackle concerns, including equity, transparency, data privacy, and security. Initiatives like as the European Union's Ethics Principles for Trustworthy AI establish a significant framework for developing ethical AI systems; nevertheless, further efforts are required to guarantee the widespread adoption and enforcement of these principles across many industries.

As AI technology advances, organisations must be cognisant of developing trends and developments in AI-driven recruitment. Prospective advancements, including AI-driven interviews, psychometric evaluations, and automated reference verification, possess the capacity to significantly alter the recruitment process. Nonetheless, these developments also elicit new ethical and practical dilemmas, especially with equity and human supervision. By anticipating these developments and consistently refining their AI tactics, organisations may guarantee their position at the vanguard of ethical and efficient recruitment procedures.

AI-driven ranking algorithms provide substantial advantages for recruitment, such as improved efficiency, augmented data-informed decision-making, and the capacity to foster workforce diversity. Nonetheless, these tools pose difficulties with algorithmic bias, transparency, privacy, and data security. To fully leverage

AI in recruitment while mitigating these problems, organisations must prioritise the creation of equitable and transparent AI models, encourage collaboration among stakeholders, and establish comprehensive legal frameworks. Subsequent studies ought to concentrate on mitigating bias, advancing inclusive recruiting methodologies, and ensuring that AI-facilitated recruitment is both pioneering and morally accountable. Organisations may develop recruitment processes that are more efficient, effective, equitable, and inclusive for all candidates.

REFERENCES

Abedi, M. (2022). *Towards a Reference Architecture of AI-Based Job Interview Systems.* Montreal, Canada: Ecole Polytechnique.

Agarwal, R., Dhar, V. (2014). Big data, data science, and analytics: The opportunity and challenge for IS research. *Inf. Syst. Res., 25*(3), 443-448.
[http://dx.doi.org/10.1287/isre.2014.0546]

Agnihotri, A., & Bhattacharya, S. (2024). Artificial Intelligence for Hiring and Induction: The Unilever Experience. SAGE Publications: SAGE Business Cases Originals..
[http://dx.doi.org/10.4135/9781071921470]

Ajunwa, I. (2021). Automated video interviewing as the new phrenology. *Berkeley Technol. Law J., 36*, 1173.

Allal-Chérif, O., Yela Aránega, A., Castaño Sánchez, R. (2021). Intelligent recruitment: How to identify, select, and retain talents from around the world using artificial intelligence. *Technol. Forecast. Soc. Change, 169*, 120822.
[http://dx.doi.org/10.1016/j.techfore.2021.120822]

Aloisi, A. (2024). Regulating algorithmic management at work in the European Union: Data protection, non-discrimination and collective rights. *Int. J. Comp. Labour Law Ind. Relat., 40*(1), 37-70.
[http://dx.doi.org/10.54648/IJCL2024001]

Andrews, L., Bucher, H. (2022). Automating Discrimination: AI Hiring Practices and Gender Inequality. *Cardozo Law Rev., 44*, 145.

Bhukya, R. (2021). Exploring Machine Learning: A Beginners Perspective. Horizon Books (A Division of Ignited Minds Edutech P Ltd).

Binns, R. (2018). Fairness in machine learning: Lessons from political philosophy. *Proceedings of the 2018 Conference on Fairness, Accountability, and Transparency,* 149-159.

Bersin, J. (2018). *AI in HR: A practical guide for HR professionals..* Josh Bersin Academy.

Bresciani, S., Ferraris, A., Romano, M., Santoro, G. (2021). Human resource management and digitalisation. *Digital transformation management for agile organizations: A compass to sail the digital world.* Emerald Publishing Limited.
[http://dx.doi.org/10.1108/978-1-80043-171-320211007]

Brynjolfsson, E., McAfee, A. (2014). *The second machine age: Work, progress, and prosperity in a time of brilliant technologies..* W. W. Norton & Company.

Charlwood, A., Guenole, N. (2022). Can HR adapt to the paradoxes of artificial intelligence? *Hum. Resour. Manage. J., 32*(4), 729-742.
[http://dx.doi.org/10.1111/1748-8583.12433]

Chen, Z. (2023). Ethics and discrimination in artificial intelligence-enabled recruitment practices. *Humanities and Social Sciences Communications, 10*(1), 1-12.

Choudhary, V., Marchetti, A., Shrestha, Y.R., Puranam, P. (2023). Human-AI ensembles: When can they work? *J. Manage.,* 01492063231194968.

Collett, C., Gomes, L.G., Neff, G. (2022). *The effects of AI on the working lives of women.*. UNESCO Publishing.
[http://dx.doi.org/10.18235/0004055]

Dastin, J. (2018). Amazon scrapped secret AI recruiting tool that showed bias against women. Reuters. Available from: https://www.reuters.com/article/us-amazon-com-jobs-automation-in-ight-idUSKCN1MK08G.

de Almeida, P.G.R., dos Santos, C.D., Farias, J.S. (2021). Artificial Intelligence Regulation: a framework for governance. *Ethics Inf. Technol., 23*(3), 505-525.
[http://dx.doi.org/10.1007/s10676-021-09593-z]

Díaz-Rodríguez, N., Del Ser, J., Coeckelbergh, M., López de Prado, M., Herrera-Viedma, E., Herrera, F. (2023). Connecting the dots in trustworthy Artificial Intelligence: From AI principles, ethics, and key requirements to responsible AI systems and regulation. *Inf. Fusion, 99*, 101896.
[http://dx.doi.org/10.1016/j.inffus.2023.101896]

Dobson, J.E. (2023). On reading and interpreting black box deep neural networks. *International Journal of Digital Humanities, 5*(2-3), 431-449.
[http://dx.doi.org/10.1007/s42803-023-00075-w]

Dwivedi, Y.K., Hughes, L., Ismagilova, E., Aarts, G., Coombs, C., Crick, T., Duan, Y., Dwivedi, R., Edwards, J., Eirug, A., Galanos, V., Ilavarasan, P.V., Janssen, M., Jones, P., Kar, A.K., Kizgin, H., Kronemann, B., Lal, B., Lucini, B., Medaglia, R., Le Meunier-FitzHugh, K., Le Meunier-FitzHugh, L.C., Misra, S., Mogaji, E., Sharma, S.K., Singh, J.B., Raghavan, V., Raman, R., Rana, N.P., Samothrakis, S., Spencer, J., Tamilmani, K., Tubadji, A., Walton, P., Williams, M.D. (2021). Artificial Intelligence (AI): Multidisciplinary perspectives on emerging challenges, opportunities, and agenda for research, practice and policy. *Int. J. Inf. Manage., 57*, 101994.
[http://dx.doi.org/10.1016/j.ijinfomgt.2019.08.002]

Garg, S., Sinha, S., Kar, A.K., Mani, M. (2022). A review of machine learning applications in human resource management. *Int. J. Prod. Perform. Manag., 71*(5), 1590-1610.
[http://dx.doi.org/10.1108/IJPPM-08-2020-0427]

Gerlings, J., Shollo, A., & Constantiou, I. (2020). Reviewing the need for explainable artificial intelligence (xAI). arXiv preprint arXiv:2012.01007.

Harinath, G. (2023). Does personal data protection matter in data protection law? A transformational model to fit in the digital era. *Handbook of Big Data Research Methods.* Edward Elgar Publishing.
[http://dx.doi.org/10.4337/9781800888555.00020]

Hassan, A. (2022). The role of artificial intelligence in human resources. *Future of Organizations and Work After the 4th Industrial Revolution: The Role of Artificial Intelligence, Big Data, Automation, and Robotics.* (pp. 57-80). Cham: Springer International Publishing.
[http://dx.doi.org/10.1007/978-3-030-99000-8_4]

Hemamou, L., Guillon, A., Martin, J.C., Clavel, C. (2023). Multimodal hierarchical attention neural network: Looking for candidatesbehaviour which impact recruiter's decision. *IEEE Trans. Affect. Comput., 14*(2), 969-985.
[http://dx.doi.org/10.1109/TAFFC.2021.3113159]

Hewage, A. (2023). Exploring the applicability of artificial intelligence in recruitment and selection processes: a focus on the recruitment phase. *J. Hum. Resour. Sustain. Stud., 11*(3), 603-634.
[http://dx.doi.org/10.4236/jhrss.2023.113034]

Hickman, E., Petrin, M. (2021). Trustworthy AI and corporate governance: the EU's ethics guidelines for trustworthy artificial intelligence from a company law perspective. *Eur. Bus. Organ. Law Rev., 22*(4), 593-625.
[http://dx.doi.org/10.1007/s40804-021-00224-0]

Hosain, M.S., Liu, P. (2020). The role of social media on talent search and acquisition: Evidence from contemporary literature. *Journal of Intercultural Management, 12*(1), 92-137. [http://dx.doi.org/10.2478/joim-2020-0034]

Jain, A., Awasar, D. (2020). Impact of artificial intelligence on recruitment: A systematic review. *Procedia Comput. Sci., 167*, 1924-1931. [http://dx.doi.org/10.1016/j.procs.2020.03.296]

Jacobs, A.M., Büthe, T., Arjona, A., Arriola, L.R., Bellin, E., Bennett, A., Björkman, L., Bleich, E., Elkins, Z., Fairfield, T., Gaikwad, N., Greitens, S.C., Hawkesworth, M., Herrera, V., Herrera, Y.M., Johnson, K.S., Karakoç, E., Koivu, K., Kreuzer, M., Lake, M., Luke, T.W., MacLean, L.M., Majic, S., Maxwell, R., Mampilly, Z., Mickey, R., Morgan, K.J., Parkinson, S.E., Parsons, C., Pearlman, W., Pollack, M.A., Posner, E., Riedl, R.B., Schatz, E., Schneider, C.Q., Schwedler, J., Shesterinina, A., Simmons, E.S., Singerman, D., Soifer, H.D., Smith, N.R., Spitzer, S., Tallberg, J., Thomson, S., Vázquez-Arroyo, A.Y., Vis, B., Wedeen, L., Williams, J.A., Wood, E.J., Yashar, D.J. (2021). The qualitative transparency deliberations: Insights and implications. *Perspect. Polit., 19*(1), 171-208. [http://dx.doi.org/10.1017/S1537592720001164]

Kabir, M. (2024). IntelCV: An Intelligent Skilled Resume Selection Method for Job Purposes (Doctoral dissertation, United International University).

Kazim, E., Koshiyama, A.S., Hilliard, A., Polle, R. (2021). Systematizing audit in algorithmic recruitment. *J. Intell., 9*(3), 46. [http://dx.doi.org/10.3390/jintelligence9030046] [PMID: 34564294]

Khan, N. The Role of Hr Professional in Technologically Advanced Hr Function. International Journal of Human Resource Management and Research (IJHRMR) 7.5, Oct 2017, 1, 8.

Khan, R., Hussain, A., Ahmad, S. (2023). Revolutionizing Human Resource Management: The Transformative Impact of Artificial Intelligence (AI) Applications. *International Journal of Social Science & Entrepreneurship, 3*(4), 306-326.

Kshetri, N. (2021). Evolving uses of artificial intelligence in human resource management in emerging economies in the global South: some preliminary evidence. *Manag. Res. Rev., 44*(7), 970-990. [http://dx.doi.org/10.1108/MRR-03-2020-0168]

Lisa, A. K., & Talla Simo, V. R. (2021). An in-depth study on the stages of AI in recruitment process of HRM and attitudes of recruiters and recruitees towards AI in Sweden.

Marr, B. (2018). How AI is changing the future of recruitment. Forbes. Available from: https://www.forbes.com/sites/bernardmarr/2018/07/02/how-ai-is-changing-the-fut-re-of-recruitment/?sh=6b11fa6b6c6f.

Peña, A., Serna, I., Morales, A., Fierrez, J., Ortega, A., Herrarte, A., Alcantara, M., Ortega-Garcia, J. (2023). Human-centric multimodal machine learning: Recent advances and testbed on AI-based recruitment. *SN Computer Science, 4*(5), 434. [http://dx.doi.org/10.1007/s42979-023-01733-0]

Pisanelli, E. (2023). Artificial intelligence, gender and work (Doctoral dissertation, European University Institute).

Pradhan, I. P., Saxena, P., Ganguly, C., & Penava, M. B. Artificial Intelligence in Recruitment: An Impact Assessment of Organizations. In Disruptive Artificial Intelligence and Sustainable Human Resource Management (pp. 123-137). River Publishers.

Raghavan, M., Barocas, S., Kleinberg, J., Levy, K. (2020). Mitigating bias in algorithmic hiring: Evaluating claims and practices. *Proceedings of the 2020 Conference on Fairness, Accountability, and Transparency,* 469-481.

Rigotti, C., Fosch-Villaronga, E. (2024). Fairness, AI & recruitment. *Comput. Law Secur. Rev., 53*, 105966. [http://dx.doi.org/10.1016/j.clsr.2024.105966]

Suneetha, S., Sudheer, A.K., Shankar, C., Metta, S. (2024). HR in the Era of AI Paradoxes: Balancing Automation and Human Touch. *International Conference on Digital Transformation in Business: Navigating the New Frontiers Beyond Boundaries (DTBNNF 2024),* 528-545.Atlantis Press.. [http://dx.doi.org/10.2991/978-94-6463-433-4_40]

Tambe, P., Cappelli, P., Yakubovich, V. (2019). Artificial intelligence in human resources management: Challenges and a path forward. *Calif. Manage. Rev., 61*(4), 15-42.

Uma, V. R., Velchamy, I., & Upadhyay, D. (2023). Recruitment analytics: Hiring in the era of artificial intelligence. In The Adoption and Effect of Artificial Intelligence on Human Resources Management, Part A (pp. 155-174). Emerald Publishing Limited.

Veglianti, E., Trombin, M., Pinna, R., De Marco, M. (2023). Customized Artificial Intelligence for Talent Recruiting: A Bias-Free Tool? *Smart Technologies for Organizations: Managing a Sustainable and Inclusive Digital Transformation.* (pp. 245-261). Cham: Springer International Publishing.

von Eschenbach, W.J. (2021). Transparency and the black box problem: Why we do not trust AI. *Philos. Technol., 34*(4), 1607-1622. [http://dx.doi.org/10.1007/s13347-021-00477-0]

Westerman, G., Bonnet, D., McAfee, A. (2014). *Leading digital: Turning technology into business transformation.* Harvard Business Review Press.

CHAPTER 11

Anticipating the Future Workforce: HR Strategies for Embracing AI and Automation

Yashashvi Jaiswal[1,*] and **Sneha Raghuvanshi**[1]

[1] *Department of Management Studies, Medi Caps University, Indore, India*

Abstract: Automation and Artificial Intelligence (AI) are redefining sectors, which will have a big impact on the workforce of the future. This book chapter examines the HR tactics that businesses need to implement to successfully adopt these new technologies. The integration of AI-driven technologies for hiring, training, and performance management is one of the main areas of concentration. Other important areas include the significance of reskilling and upskilling workers to match changing job needs and the function of change management in promoting an innovative and adaptable culture. This book chapter offers a thorough roadmap for HR professionals to handle the possibilities and difficulties brought about by AI and automation, guaranteeing a resilient and future-ready workforce, by looking at best practices and case studies. Human Resources (HR) departments are leading this shift as automation and Artificial Intelligence (AI) revolutionize businesses and employment roles. This paper explores the strategic measures that HR must take to fully utilise automation and artificial intelligence while guaranteeing a smooth transition into the workforce. It starts with a review of automation and Artificial Intelligence (AI) technologies, emphasizing their potential and the changes they can bring to conventional employment functions. After that, the emphasis switches to AI-driven hiring and talent acquisition, investigating cutting-edge technologies that improve the processes of finding, vetting, and choosing candidates while taking ethical issues into account.

Keywords: Adaptability, AI integration, Automation, Change management, Continuous learning, Cybersecurity, Data-driven HR, Digital competence, Digital workforce, Employee engagement, Employee experience, Future of work, Human-AI collaboration, Hybrid workforce, Reskilling, Smart workplaces, Strategic HR, Talent acquisition, Talent management, Upskilling.

INTRODUCTION

Automation and Artificial Intelligence (AI) are becoming a reality in the workplace and are changing sectors all over the world. They are no longer just

* **Corresponding author Yashashvi Jaiswal:** Department of Management Studies, Medi Caps University, Indore, India; E-mail: yashasvijaiswal0509@gmail.com, drsnehajadonraghuvanshi@gmail.com

Rishikaysh Kaakandikar, Keshav Kaushik, Supriya Pathakm Madhavi Shamkuwar & Vivek Mishra (Eds.)

theoretical concepts. AI and automation are boosting productivity, streamlining procedures, and upending conventional business models in a variety of industries, including manufacturing, banking, healthcare, and retail. These technologies will unavoidably and profoundly affect the workforce as they develop. Work jobs are changing; some are going away, and new possibilities are forming that call for various skill sets and methods of operation. Human Resources (HR) is playing a more strategic role in this new era. The difficult job of managing this shift falls to HR specialists, who must make sure that businesses can take full advantage of automation and artificial intelligence while simultaneously attending to the needs of people. To achieve this, traditional HR responsibilities must give way to a more proactive, forward-thinking strategy that prepares for change and develops a staff that is resilient and adaptive. It takes more than just implementing new technology to embrace AI and automation in the workplace; it involves reevaluating every step of the employee lifecycle, from hiring and onboarding to talent development and retention. Hiring procedures are being revolutionized by AI-powered technologies that enhance applicant matching and lessen prejudice. Routine administrative chores are becoming more efficient thanks to automation, which frees up HR specialists to concentrate on more strategic projects. But these developments also mean that workers need to learn new skills, especially in data literacy, digital fluency, and human-AI cooperation. Since present abilities are likely to become outdated due to rapid technological advancements, the workforce of the future will need to be flexible and always learning new things. Prioritizing reskilling and upskilling efforts is crucial for HR directors to guarantee that workers stay relevant and have the ability to prosper in an automated future. For businesses to remain competitive and adaptable to changes in the market, they must also cultivate an innovative and agile culture. Ethical issues need to be foremost when HR practices change to incorporate AI and automation. HR professionals face several important problems, including ensuring equitable treatment of employees, resolving concerns about job displacement, and managing the ethical use of AI. The secret to building a sustainable and diverse workforce in the future will be to strike a balance between the efficiency advantages of technology and the necessity of meaningful human labor. The groundwork for comprehending the revolutionary effects of automation and artificial intelligence on the workforce, as well as the crucial role HR strategies play in managing this transition, is laid out in this introduction. HR directors can help their companies navigate the challenges of the digital era and ensure that business goals and employee welfare are in sync with the shift to a digital workforce by carefully adopting these technologies.

AI technologies have increasingly become essential in transforming HR practices, including recruitment and performance management (Tambe *et al.*, 2019; Davenport & Ronanki, 2018). According to Bessé (2019), AI can impact jobs by reshaping demand, and researchers like Guenole *et al.* (2017) argue that the business case for AI in HR lies in its ability to automate learning and development processes. However, challenges remain, particularly in ensuring fairness and mitigating biases in AI algorithms (Binns, 2018; Raghavan *et al.*, 2020). AI's role in HR has also led to concerns regarding surveillance and privacy, as explored by Moore (2019), who highlights the evolving intersection of technology and workplace rights. The strategic integration of AI can significantly enhance decision-making, as emphasized by Bughin *et al.* (2018) and Westerman *et al.* (2014). Despite its potential, experts such as Cappelli (2019) and Pulakos and O'Leary (2011) argue that traditional HR practices must adapt to leverage AI effectively, while others, including Cowgill *et al.* (2020), suggest that bias in data or programming can hinder AI's efficacy in the HR context.

SCOPE OF AI IN HUMAN RESOURCES

Artificial Intelligence (AI) is transforming Human Resources (HR) by improving decision-making, automating processes, and streamlining workforce management across HR operations. Artificial intelligence (AI)-driven solutions in recruiting expedite the screening of candidates, boost chatbot interaction, and improve hiring forecasts. AI streamlines onboarding procedures and provides virtual assistants to assist recent hires. AI detects skill gaps and develops individualized training programs for learning and growth. By lowering evaluation biases and offering ongoing feedback, it also helps with performance management. Personalized wellness programs enhance employee well-being, while AI-driven analytics, through sentiment analysis and predictive models, assist in boosting employee engagement and retention. AI is essential for HR analytics and decision-making as well, providing insights into labour patterns and helping to maximize benefit packages. Through talent pool expansion and the detection of biases in HR procedures, it fosters diversity and inclusion. Additionally, by spotting possible workplace problems, AI controls risks and aids in maintaining compliance with labor rules. AI enables employee self-service and automates repetitive administrative duties, freeing HR experts to concentrate on strategic projects. All things considered, integrating AI into HR improves productivity, encourages better decision-making, and gets the workforce ready for the future; yet, its adoption requires careful planning and ethical concerns.

HISTORICAL CONTEXT:

How AI has evolved and its entry into HR: The history of Artificial Intelligence (AI) is lengthy and dynamic; it started out as a theoretical idea before progressively becoming a crucial component of many different businesses, including Human Resources (HR). Knowing the background of AI and how it entered the HR field will help one better understand how it changed from being a future concept to a useful tool that is transforming the workforce. Early in AI Development, AI's origins may be found in the middle of the 20th century, when early studies concentrated on the prospect of building robots that could simulate the human mind. At the Dartmouth Conference in 1956, the term "artificial intelligence" was formally created, establishing AI as a separate area of research. In the decades that followed, Artificial Intelligence (AI) research concentrated on logic, reasoning, and problem-solving, albeit technological limitations prevented many real-world applications. With the advent of expert systems—software intended to mimic human decision-making in particular fields—AI saw a rebirth in the 1980s. These early AI systems lacked the flexibility and learning capacities of contemporary AI; instead, they were rule-based and dependent on large knowledge stores for decision-making.

Artificial Intelligence's Ingress into Human Resources

The early 2010s saw the start of AI's incorporation into HR as businesses saw how much AI could automate and improve a range of HR processes. When AI was first introduced into HR, it was primarily utilized to automate labor-intensive processes like arranging interviews and screening resumes, especially for talent acquisition and recruiting. AI-powered technologies were created to go through a lot of applications and find eligible applicants based on preset standards, saving time and effort compared to manual screening. AI technology's uses in HR grew as it developed. HR was able to make more data-driven decisions in areas like hiring, performance management, and employee retention with the help of predictive analytics, which developed into a potent tool. AI may examine past data to find trends and forecast future results, assisting HR professionals in making more strategic and well-informed decisions. With the development of Natural Language Processing (NLP) and AI-powered chatbots, the use of AI in HR has increased even further. By offering real-time help and automating repetitive enquiries, these technologies improved employee engagement and freed up HR personnel to concentrate on higher-value work. AI-driven systems also started to help with staff development, training, and onboarding by providing individualized learning experiences that were catered to each worker's needs.

ARTIFICIAL INTELLIGENCE IN HUMAN RESOURCES

AI is becoming a crucial part of contemporary HR procedures, revolutionizing the way businesses handle their personnel. AI is assisting HR departments in increasing productivity, minimizing biases, and improving employee experiences through anything from AI-powered recruiting tools to customized employee development programs. These days, the whole employee lifecycle—from hiring and onboarding to performance management and retention—is integrated with technology.

Additionally, HR practitioners may obtain better insights into worker patterns and behaviors because of AI's capacity to analyses massive volumes of data, which makes proactive and strategic HR management possible. For example, sentiment analysis powered by AI may assist businesses in determining employee engagement and spotting any problems before they become serious.

Although AI's introduction into HR has many positive effects, it has also brought up ethical issues such as data privacy, openness, and algorithmic prejudice. HR specialists need to carefully manage these difficulties to guarantee that AI is applied in a responsible and inclusive manner.

LITERATURE REVIEW

HR strategies must adapt to the changing nature of work, where AI and automation are replacing routine tasks and creating demand for new skills. One of the primary challenges is workforce planning, which involves predicting the skills and competencies required in the future. According to Bessen (2019), automation is not simply eliminating jobs but transforming them, requiring HR professionals to anticipate the changing nature of job roles and recruit talent with the necessary skills. Additionally, Cappelli (2019) highlights that AI can streamline the recruitment process by using predictive analytics to identify top candidates more efficiently.

The use of AI in workforce planning allows HR departments to optimize talent acquisition by analyzing data on industry trends, skill shortages, and workforce demographics. As noted by Tambe, Cappelli, and Yakubovich (2019), predictive analytics can help organizations anticipate future workforce needs and develop proactive hiring strategies. This approach is particularly valuable in industries undergoing rapid technological change, where traditional hiring methods may fall short in identifying future skill requirements.

AI and automation are also transforming how HR departments approach employee engagement and retention. AI-driven tools can provide real-time insights into employee behavior, enabling HR to proactively address issues that might lead to dissatisfaction or turnover. According to Bersin (2018), AI can be used to monitor employee engagement levels and suggest personalized interventions to improve job satisfaction and productivity.

Moreover, automation can reduce administrative burdens on HR professionals, allowing them to focus on more strategic activities such as employee engagement.

As Wright and Ulrich (2017) suggest, by automating routine tasks, HR can devote more time to developing programs that enhance employee experience and foster a positive work environment. This shift not only improves employee satisfaction but also helps in retaining top talent in a competitive labor market.

As AI and automation continue to evolve, continuous skill development will be critical for the workforce to stay relevant. HR strategies must emphasize upskilling and reskilling employees to meet the demands of a more technologically driven workplace. According to Bughin *et al.* (2018), as automation displaces certain tasks, the demand for advanced technological skills, as well as soft skills like creativity and problem-solving, will increase.

HR can leverage AI to personalize learning and development programs, tailoring them to individual employee needs and career goals. AI-driven learning platforms can assess employees' current skill levels and recommend targeted training programs to help them acquire new competencies. This approach not only enhances the effectiveness of training initiatives but also ensures that the workforce is continuously adapting to technological advancements (Westerman, Bonnet, & McAfee, 2014).

Despite the potential benefits, the integration of AI and automation into HR processes raises several ethical concerns. The use of AI in decision-making, particularly in hiring and performance evaluations, may introduce bias if not properly managed. As Binns (2018) notes, AI systems can unintentionally perpetuate existing biases in the data they are trained on, leading to discriminatory outcomes. HR professionals must ensure that AI tools are designed and implemented with fairness and transparency in mind, involving regular audits to detect and mitigate bias.

Additionally, the increased use of AI and automation in the workplace necessitates robust data privacy and security measures. HR departments handle vast amounts of sensitive employee data, and the integration of AI systems requires a careful balance between leveraging this data for insights and protecting employee privacy (Moore, 2019). Ensuring compliance with data protection regulations and implementing strong cybersecurity protocols are essential for maintaining trust in AI-driven HR initiatives.

AI and automation in HR can help improve diversity and inclusion by reducing bias in recruitment and performance management processes. Research shows that AI-driven recruitment tools can eliminate human biases from initial candidate screening by focusing solely on skill-based criteria (Cowgill, Dell'Acqua, & Deng, 2020). This can lead to a more diverse pool of candidates and help organizations meet their diversity goals.

However, there are concerns that AI systems, if not properly monitored, can unintentionally perpetuate existing biases found in historical data. Binns (2018) emphasizes the need for HR departments to actively audit AI algorithms to prevent bias, ensuring that diversity and inclusion remain central in AI-driven processes. Additionally, companies must continuously update and refine their data sets to avoid reinforcing outdated or discriminatory hiring patterns (Raghavan *et al.*, 2020).

AI and automation also play a significant role in performance management by providing real-time data and insights into employee productivity. Automated performance tracking tools can help HR professionals identify top performers and those in need of development (Bersin, 2018). According to Pulakos and O'Leary (2011), AI can assist in creating more objective and data-driven performance evaluations, reducing the subjectivity that often accompanies traditional methods.

However, AI-based performance management also requires transparency. Employees need to understand how their performance is being evaluated by AI systems, and HR must ensure that these systems are fair and aligned with the organization's values (Bersin, 2018). Moreover, automated systems should complement, rather than replace, the human touch in performance discussions, as feedback from managers and peers remains essential for employee development (Tambe, Cappelli, & Yakubovich, 2019).

The introduction of AI and automation into HR necessitates a broader cultural transformation within organizations. AI-driven systems require a shift in how employees perceive their roles and the value they bring to the organization. According to Davenport and Ronanki (2018), successful integration of AI in HR requires a culture of continuous learning, where employees are encouraged to embrace new technologies and develop new skills.

Moreover, change management strategies are crucial to ensuring a smooth transition to AI-driven HR practices. HR leaders must communicate the benefits of AI and automation to the workforce and provide adequate training to help employees adapt (Westerman, Bonnet, & McAfee, 2014). Resistance to AI-driven changes can be mitigated through transparent communication and by involving employees in the decision-making process, thereby fostering a sense of ownership over the transformation (Bughin *et al.*, 2018).

AI and automation are also reshaping leadership development programs by identifying potential leaders and personalizing development opportunities. AI can analyze employee data to uncover leadership potential that might otherwise be overlooked through traditional assessment methods (Guenole, Ferrar, & Feinzig, 2017). According to Avolio, Sosik, and Berson (2013), AI-driven insights can help HR professionals tailor leadership development programs to the specific needs of emerging leaders, increasing the effectiveness of these initiatives.

Workforce analytics powered by AI allows HR departments to make more data-driven decisions regarding employee engagement, retention, and productivity. AI can provide insights into patterns and trends that may not be immediately apparent through traditional HR metrics (Bersin, 2018). For example, predictive analytics can help HR anticipate turnover risks by identifying factors that contribute to employee dissatisfaction (Guenole *et al.*, 2017).

However, as with other AI-driven HR processes, the use of workforce analytics raises ethical considerations related to employee privacy. HR professionals must ensure that data is used responsibly and that employees are informed about how their data will be used (Moore, 2019). Ensuring that AI-driven workforce analytics are aligned with ethical standards is crucial to maintaining trust and protecting employee rights (Manyika *et al.*, 2017).

ARTIFICIAL INTELLIGENCE IN TALENT ACQUISITION

Through the automation of processes, improvement of decision-making, and enhancement of the candidate experience, Artificial Intelligence (AI) is profoundly changing the talent acquisition landscape. AI-driven platforms and solutions are revolutionizing the recruiting process, from candidate sourcing to onboarding, by increasing speed, accuracy, and efficiency. The following are the main ways that AI is affecting hiring:

AUTOMATED RESUME SCREENING

Efficiency Gains: Going through resumes is one of the most time-consuming parts of hiring. AI-powered systems can swiftly analyze and sort applications, finding the best applicants based on preset standards like education, experience, and abilities. Recruiters' manual effort is greatly reduced by this automation, freeing them up to concentrate on more important duties.

Reduced Bias: By programming AI systems to assess applicants only based on objective standards, unintentional biases that may arise during human screening can be lessened. Artificial Intelligence enables a more equitable and inclusive first selection process by anonymizing applicant data.

Example: Hiring Vue and Pymetrics are two companies that utilize AI to automatically filter resumes by comparing them to job descriptions through the analysis of keywords, skills, and experience. Recruiters may now concentrate on the best applicants instead of having to go through hundreds of resumes by hand.

CANDIDATE SOURCING ENGAGEMENT

Proactive Sourcing: AI systems can search *via* a variety of channels (such as social media and job boards) and find passive applicants who fit the job specifications. Recruiters may increase the pool of talent they have access to by using AI to evaluate profiles and recommend individuals who might not have applied but are a good fit.

Chatbots for Engagement: During the hiring process, chatbots driven by artificial intelligence are being utilized more and more to interact with candidates. Chatbots may set up interviews, schedule them, and respond to often-requested enquiries about the position and business. This degree of automation shortens wait times and gives instantaneous replies, improving the applicant experience.

Example: Using AI, LinkedIn Talent Insights finds and suggests passive applicants based on job needs. To further improve the applicant experience, Mya Systems offers an AI chatbot that converses with candidates, responding to enquiries and setting up interviews.

PREDICTIVE ANALYTICS FOR HIRING DECISIONS

Enhanced Decision-Making: AI-powered predictive analytics can anticipate a candidate's likelihood of success in a position by analyzing their past data, abilities, and behavioral patterns. With the use of data, recruiters can make better judgments, lowering the possibility of unsatisfactory hires and raising the standard of recruiting all around.

Cultural Fit Analysis: By examining a candidate's behavior, values, and communication style, AI can determine whether or not they will mesh well with the company's culture. This guarantees that recent recruits are compatible with the organization's values in addition to their skill set, which may increase retention rates.

Example: Predictive Hire analyzes interviewees' replies, psychometric data, and prior experiences to determine how well they will do on the job using AI-driven predictive analytics. This predicts a candidate's chances of succeeding in a post and helps organizations make better recruiting decisions.

INTERVIEW PROCESS OPTIMIZATION

Automated Video Interviews: Video interviews are one of the ways through which artificial intelligence is being utilized throughout the interview process. To evaluate a candidate's emotional intelligence, stress tolerance, and general communication abilities, AI can examine their facial expressions, voice tones, and

speech patterns during interviews. Recruiters may assess candidates further with the use of these insights, which go beyond conventional techniques.

Interview Scheduling: AI-powered systems are capable of handling interview scheduling on their own. They can coordinate with interviewers and applicants to determine the most convenient times, which eliminates administrative work and back-and-forth communication.

Example: HireVue provides video interviews with AI-powered analysis of candidates' language, tone of voice, and facial expressions. Recruiters gain further insights by using the system's applicant scoring, which is based on these criteria. By matching the calendars of recruiters and applicants, XOR automates the process of organizing interviews by determining the most convenient times.

DIVERSITY AND INCLUSION IN HIRING

Bias Reduction: By emphasizing abilities and credentials above variables like gender, race, or age, AI can help lessen prejudices in the hiring process. Certain AI technologies are made expressly to evaluate job descriptions and eliminate discriminatory wording, encouraging a more diverse recruiting process.

Diverse Talent Pools: By extending the search parameters outside conventional networks, Artificial Intelligence (AI) may be utilized to find and connect with individuals from underrepresented groups. This proactive strategy aids businesses in developing a more diversified workforce.

Example: Using AI, Textio evaluates job descriptions and makes recommendations for modifying them to remove discriminatory wording that might turn off individuals from various backgrounds. Entelo is an additional AI tool that expands search parameters beyond standard networks and uses machine learning to assist organizations in locating diverse individuals.

CANDIDATE EXPERIENCE ENHANCEMENT

Customized Interactions: By adjusting communication and information to each candidate specifically, artificial intelligence may make the hiring process more unique. AI may, for instance, deliver tailored messages or suggest jobs based on a candidate's profile, increasing engagement and piquing their interest all the way through the recruiting process.

Real-Time Feedback: AI-powered systems can give applicants feedback in real-time, updating them on the status of their applications and the next stages. The applicant experience is enhanced overall, and irritation from protracted wait times is decreased because of this openness.

Example: Beamery uses AI to personalize communication with candidates, offering tailored job recommendations and updates throughout the recruitment process. Olivia by Paradox is an AI assistant that provides real-time feedback to candidates, keeping them engaged and informed about their application status.

ONBOARDING AND INTEGRATIONS

Automated Onboarding: AI is used in the onboarding process to automate administrative labor, including arranging orientations, completing paperwork, and enrolling in benefits. To guarantee a seamless and uniform onboarding process, AI-powered onboarding solutions may also lead new hires through corporate rules, training materials, and role-specific resources.

Long-Term Retention Predictions: AI can also help with onboarding by identifying new recruits who have the highest likelihood of sticking with the firm over the long run. This information helps HR teams to better support and assist new hires throughout the crucial initial phases of their employment.

Example: Sapling is an AI-driven onboarding software that streamlines administrative duties such as arranging orientation, enrolling in benefits, and completing paperwork. By giving new recruits pertinent resources and knowledge suited to their roles, Enboarder's AI-powered personalized onboarding experiences facilitate a smoother transition into the organization.

DATA DRIVEN INSIGHTS FOR CONTINUOUS IMPROVEMENT

Recruitment Analytics: AI-driven analytics tools offer insights into the effectiveness of recruitment strategies by tracking key metrics such as time-t--hire, cost-per-hire, and candidate satisfaction. These insights enable HR teams to optimize their talent acquisition processes continually.

Trend Analysis: AI can analyze trends in the job market, such as shifts in candidate expectations, emerging skill sets, or salary benchmarks, allowing companies to adapt their talent acquisition strategies proactively to stay competitive.

Example: Hiring parameters, including time-to-hire, cost-per-hire, and quality of hiring, are tracked by Visier's AI-driven analytics. HR departments may use these findings to pinpoint hiring process bottlenecks and gradually improve their tactics. Recruiters may keep abreast of industry changes and modify their talent acquisition strategy with the aid of Smart Recruiters' AI-powered trend analysis.

EMPLOYEE ONBOARDING AND INTEGRATION

Artificial Intelligence has the potential to greatly improve HR's employee onboarding and integration procedures by optimizing workflows, enhancing customization, and guaranteeing a more seamless onboarding experience for new hires. Here are a few ways AI may help:

AUTOMATING TASKS RELATED TO ADMINISTRATION

Document Management: AI can create, send, and track onboarding papers like tax forms, contracts, and NDAs automatically. It may verify that all documentation is complete and alert HR to those that are lacking.

Task Scheduling: HR staff may eliminate the need for manual scheduling by using AI-powered solutions to set up and manage calendars for onboarding meetings, training sessions, and check-ins.

Compliance: By keeping track of and reminding staff members of deadlines, AI may assist in making sure that all required compliance training is completed.

TAILORED INDUCTION PROCESSES

Tailored Learning Paths: AI may develop individualized onboarding programs by evaluating a new hire's job, experience, and historical performance. It can also suggest mentors, resources, and training courses.

Chatbots and virtual assistants: AI-driven chatbots may provide round-th--clock answers to frequently asked onboarding enquiries, assisting new hires throughout their initial days of work without requiring continual human assistance.

Comments Gathering: AI can automatically collect and recommend improvements based on new recruits' comments on their onboarding experience.

SOCIALIZATION AND INTEGRATION

Link to Peers: AI may recommend links with coworkers based on responsibilities, hobbies, and abilities, encouraging teamwork and integration. Additionally, it might suggest networks or workplace organizations that new hires should join.

Cultural Integration: AI can evaluate a new hire's level of comprehension of the corporate culture and provide them with tools and advice to assist them in adjusting more successfully.

Mentorship Pairing: AI-powered mentorship pairing allows for more individualized and fruitful mentoring relationships by matching up new hires with mentors according to their interests, skill gaps, and career objectives.

TRACKING DEVELOPMENT AND INVOLVEMENT

Performance Tracking: During the onboarding process, AI may monitor a new hire's development by examining performance information, engagement levels, and feedback to spot problems and provide solutions when necessary.

Engagement Tools: By providing gamified onboarding procedures, immediate recognition, and continuous assistance, AI-powered systems can maintain new hire engagement.

Monitoring of Wellbeing: Artificial Intelligence (AI) systems may use sentiment analysis to analyses the emotions of new hires and determine whether they require extra assistance or resources while they adjust.

MAKING DECISIONS BASED ON DATA

Predictive Analytics: AI can evaluate information from past onboarding procedures to identify the elements that are most important for a smooth onboarding experience and employee retention. This allows HR to iteratively enhance and optimize its procedures.

Trend Analysis: HR departments may be proactive in addressing possible areas of concern and trends in onboarding success by using AI.

Customized Reporting: AI can produce comprehensive reports for management that emphasize important parameters like overall satisfaction, retention rates, and time-to-productivity during the onboarding process.

LEARNING AND DEVELOPMENT

Customized Development Plans and Learning Pathways

AI for Personalization: AI may provide a tailored learning path based on an employee's historical performance, career objectives, and present skill set. According to their requirements and goals, it suggests certain training courses, programs, and resources. For example, AI may recommend management classes or leadership seminars for an individual who demonstrates leadership aptitude but lacks formal training.

Predictive analytics and adaptive learning technologies

Adaptive Learning: Based on the learner's progress and performance, AI-driven adaptive learning systems modify the content's kind and complexity. This implies that the system may give more sophisticated content and extra help when necessary if an employee shows excellence in particular areas.

Predictive analytics is a tool that Artificial Intelligence (AI) utilizes to find possible skill gaps before they become issues. Artificial Intelligence (AI) may predict future training needs and provide development strategies to proactively fill these gaps by analyzing trends and performance data.

Assessing the Effectiveness of Training

Training Evaluation: AI technologies may monitor a variety of variables, including employee performance gains, skill acquisition rates, and participant feedback, in order to evaluate the efficacy of training initiatives. This makes it easier to assess whether training goals are being accomplished and whether any modifications are required.

Continuous Improvement: AI is able to assess the success rates of various training regimens and educational materials, giving valuable information about the most efficient techniques. This enables businesses to improve results by iteratively improving their learning and development initiatives.

PERFORMANCE MANAGEMENT AND EMPLOYEE ENGAGEMENT

Constant Evaluation and Feedback

Real-Time input: AI-powered systems can continually gather and examine user input, giving staff members instantaneous performance insights. Compared to customary quarterly or annual assessments, this facilitates the quicker resolution of problems and acknowledgement of accomplishments.

Performance Analytics: AI can analyze performance data to find patterns, advantages, and opportunities for development. It may monitor Key Performance Indicators (KPIs) and provide information on how staff members are doing in relation to predetermined targets.

ESTABLISHING AND MONITORING GOALS

Aligning objectives: AI may assist in helping employees define SMART (Specific, Measurable, Achievable, Relevant, Time-bound) objectives according to their jobs and previous performance. It guarantees that these aims coincide with the purposes of the organization.

Progress tracking: AI systems are able to keep track of how well objectives are being met and may offer regular reports as well as modification ideas. This aids in keeping things in perspective and tackling problems early.

REDUCING BIAS AND EQUITABLE ASSESSMENTS

Data-Driven assessments: By using AI to objectively analyses performance data, assessments may be less likely to be biased by humans. It guarantees that objective measurements, as opposed to subjective judgements, form the basis of performance appraisals.

Skill Gap Analysis: Using performance data, AI can pinpoint areas where employees lack certain abilities and provide tailored development programs to help them grow professionally.

SENTIMENT EVALUATION AND COMMENTARY

Employee Sentiment: AI systems are able to assess overall sentiment and engagement levels by analyzing employee comments, surveys, and messaging. This aids in assessing worker morale and resolving any problems that can affect involvement.

Pulse Surveys: AI-powered pulse surveys are brief questionnaires that are sent out on a frequent basis to gauge user involvement and gather input on modifications or issues. This enables prompt resolution of new issues as they arise.

Tailored Recommendations: AI can provide initiatives and tactics for individualized employee engagement based on their preferences and actions. For instance, it can recommend professional development possibilities catered to each person's requirements, team-building exercises, or recognition programs.

Effective Communication: AI can examine communication trends and recommend the most effective channels or times to communicate with staff members.

Behavioral Insights: AI may give businesses information on the preferences and behavior of their employees, enabling them to better target their engagement campaigns.

HR ANALYTICS AND DECISION MAKING

HR analytics is the process of measuring and comprehending worker performance, behaviour, and results using data analysis and statistical models. It is also referred to as people analytics, workforce analytics, or talent analytics. Enhancing decision-making in areas like hiring, staff retention, performance management, and workforce planning overall is the goal. The evolution of HR from a mostly administrative function to a strategic one depends heavily on HR analytics. HR workers may make better decisions that align with the company's goals by using data.

- Increased Decision Accuracy: When making judgments based on data rather than just intuition, judgments are made with greater objectivity and are less likely to be biased.
- Proactive HR Management: HR departments may take proactive measures by using predictive analytics to foresee issues like employee turnover or talent shortages.
- Improved employee happiness: By using data to better understand the needs and behaviors of employees, engagement tactics may be improved, which in turn improves retention and satisfaction.
- Alignment with Business Strategy: By maximizing talent management and workforce productivity, HR analytics assist HR departments in supporting larger business objectives.

HUMAN RESOURCE TOOLS AND TECHNOLOGIES

Numerous technologies and solutions that make data gathering, processing, and reporting easier are used to support HR analytics. Important instruments consist of:

- HR Information Systems (HRIS): These systems automate payroll, benefits administration, and recruiting while centralizing employee data.
- Business Intelligence (BI) Tools: Reports, dashboards, and visualizations that offer insights on labour patterns are created using tools like as Tableau, Power BI, and Qlik Sense.
- Artificial Intelligence and Machine Learning Platforms: AI-powered platforms assist with predictive analytics by examining trends in personnel data to anticipate outcomes like performance or attrition.
- Survey Tools: Employers may use Qualtrics or SurveyMonkey to collect employee input, which can then be analysed to determine levels of satisfaction and engagement.

EMPLOYEE EXPERIENCE AND WELL-BEING

Modern HR methods now centre on employee experience and well-being, with an emphasis on making sure that workers feel encouraged, involved, and content in their positions. Artificial Intelligence (AI) is becoming more and more crucial in tracking and improving worker well-being, tailoring assistance to individual workers, and maximizing benefits. The following are some significant ways AI is changing the work environment:

AI IN MONITORING AND ENHANCING EMPLOYEE WELL-BEING

HR departments can now more effectively monitor employee well-being and take proactive measures to address concerns as they emerge, thanks to AI-powered solutions.

Real-Time Monitoring: AI can identify patterns that can point to stress, burnout, or disengagement by analyzing data from a variety of sources, including emails, productivity tools, and employee surveys. Natural Language Processing (NLP), for example, can evaluate the communication's tone to spot indications of annoyance or discontent. HR may be notified when worrying trends are found so that the proper steps, such as modifying workloads, can be taken.

Wearable Technology: Heart rate, sleep habits, and physical activity are just a few of the parameters that AI-driven wearables can track to keep an eye on physical well-being. This information can shed light on the health of the workforce and recommend actions to enhance their general well-being, such as increasing physical activity or healthier sleeping patterns. Wellness programs for the entire organization can also be informed by anonymized and aggregated wearable data.

Sentiment Analysis: Artificial intelligence is capable of analyzing employee input to uncover underlying feelings and issues. By doing this, HR can learn more about the overall attitude of the company and deal with problems before they get out of hand. AI is also capable of analyzing feedback patterns over time, which aids HR in comprehending the effects of various efforts on well-being.

CHATBOTS AND VIRTUAL ASSISTANTS FOR EMPLOYEE SUPPORT

AI-powered chatbots and virtual assistants are becoming a crucial part of employee support since they provide information and help in real time, improving employee satisfaction while lightening the burden on HR staff.

Availability Around-the-Clock: Chatbots can give workers prompt responses to frequently asked HR-related concerns, such as those concerning payroll, benefits, or leave regulations. By enabling employees to access information whenever they need it and eliminating the need to wait for HR professionals during business hours, this can increase employee satisfaction.

Personalized Assistance: By comprehending the unique requirements of every employee, sophisticated AI-driven chatbots are able to provide individualized help. For instance, a chatbot may help a worker who has voiced worries about mental health discover appropriate resources or walk them through the process of signing up for a wellness program that suits their interests.

Support for Onboarding: Chatbots may offer new hires a virtual onboarding experience, guiding them through corporate regulations, facilitating connections with coworkers, and educating them about company culture. Chatbots may improve the first employee experience by streamlining and de-stressing onboarding.

Coaching and Development: Artificial Intelligence (AI)-enabled virtual assistants can also serve as digital coaches, offering staff members tailored suggestions for professional growth. These suggestions might include courses, mentorship programs, or skill-building exercises based on performance data and career objectives.

AI-DRIVEN PERSONALIZATION OF EMPLOYEE BENEFITS

AI is changing the way businesses create and provide benefits to their employees, making them more individualized and tailored to each worker's requirements.

Tailored Benefits Packages: These packages are specifically tailored to each employee's requirements and preferences can be created using AI's analysis of employee data. Younger workers could priorities student loan help, but older workers might give retirement planning or healthcare perks more importance. AI can assist HR in creating adaptable benefit plans that entice a varied workforce.

Predictive perks: Using information from past employee data, behaviors, and life phases, AI can forecast which perks will be most useful to certain workers. For instance, AI can recommend family-friendly perks like daycare assistance or parental leave alternatives to an employee who recently became a parent. Benefits are proactively delivered at the appropriate moment and are relevant, thanks to predictive analytics.

Employee Wellness Programs: By examining employee health information, preferences, and participation patterns, AI can tailor wellness initiatives. For instance, AI may recommend increasing access to mental health or health coaching services if an employee has been making use of fitness-related perks. AI can improve program efficacy and engagement by customizing wellness initiatives to each participant's requirements.

Financial Wellness: AI has the potential to help staff members make wiser financial decisions. For example, AI-powered applications may assess a worker's debt levels, retirement savings, and spending habits to offer tailored financial advise, including recommending changes to the budget or maximizing 401(k) contributions. AI-powered financial wellness initiatives can assist staff members in reaching their financial objectives, lowering stress levels, and improving general well-being.

AUTOMATION OF ROUTINE HUMAN RESOURCE TASKS

The use of AI and Robotic Process Automation (RPA) to automate routine HR processes is revolutionizing the way HR departments work by freeing them up to concentrate on more strategic projects rather than administrative duties. HR departments may increase overall efficiency, minimize administrative work, and simplify operations by integrating AI and RPA.

STREAMLINE HR OPERATIONS WITH AI AND RPA

Onboarding: RPA may handle the gathering of documents, completing forms, and setting up orientation sessions to automate the onboarding process. AI-driven virtual assistants may aid with onboarding, provide information, and answer questions for new hires. This guarantees a seamless and effective onboarding process with the least amount of manual involvement.

Payroll Processing: Among HR's most tedious and time-consuming responsibilities is payroll processing. Payroll procedures like benefit administration, tax deductions, and salary computations may be automated with RPA. By spotting irregularities in payroll data, such as erroneous deductions or overpayments, AI may help ensure accuracy by lowering mistakes and guaranteeing compliance.

Time and Attendance Management: By extracting information from time-keeping systems and incorporating it into payroll and HR systems, RPA can automate time tracking and attendance management. This lowers the possibility of mistakes and does away with human input. AI may also examine attendance data to spot patterns like persistent tardiness or absence, giving HR the opportunity to proactively handle such problems.

Management of Personnel Records: Keeping up with personnel records is sometimes a labor-intensive undertaking. RPA can update and extract data from several systems, automating the maintenance of these records. RPA, for example, may automatically update personnel data when an employee changes their residence or marital status, or when they take on new responsibilities inside the company.

Compliance and Reporting: By automating the creation of reports needed for internal reviews, legal filings, and audits, AI and RPA may assist HR departments in adhering to regulatory requirements. By ensuring that reports are timely and correct, these technologies help lower the chance of non-compliance and penalties.

Reducing Administrative Burden through AI-Powered Automation: AI and RPA may greatly lessen the administrative load on HR professionals by automating repetitive operations. This frees them up to concentrate on more strategic projects like workforce planning, employee engagement, and talent development.

Employee Queries and Support: AI-powered chatbots may assist with common employee enquiries like leave management, perks, and HR procedures. As a result, there are fewer queries that HR managers must manually respond to. Chatbots may be used around the clock to respond to workers instantly, increasing employee happiness and freeing up HR personnel to work on more difficult projects.

Performance Management: Gathering performance information from various systems and putting it into performance reviews may be automated with RPA. AI can analyze performance data to reveal information about an employee's areas of strength, growth, and career advancement possibilities. By automating these procedures, managers may concentrate on coaching and development while reducing the administrative burden related to performance management.

Employee Exit Processes: Automating the leaving process for employees may simplify offboarding activities, including collecting corporate assets, deactivating accounts, and processing last payments. HR personnel will have less administrative work to do and can ensure that all required procedures are performed during offboarding by using RPA to handle these activities consistently and effectively.

Distribution and Analysis of Surveys: AI and RPA can automate employee surveys, such as engagement or satisfaction surveys, in terms of distribution and analysis. While AI may analyses survey responses to find important patterns and insights, RPA can manage the logistics of getting the surveys to the right personnel. Instead of wasting time on manual survey administration, HR can now concentrate on creating plans to address employee problems thanks to automation.

CHALLENGES AND LIMITATION OF AI IN HR

AI in HR has several advantages but also presents several difficulties, such as:

Discrimination and Bias: AI may carry over preconceived notions from data, which might result in unjust employment or promotion choices.

Lack of Transparency: Decision-making is tainted by the difficulty of interpreting AI models. Data Security and Privacy: Managing private employee information gives rise to privacy and security issues.

Job Displacement: AI automation may raise concerns about losing one's job, which might lower morale.

Implementation Complexity: It can be difficult and resource-intensive to integrate AI into HR systems.

Limited Understanding: When making decisions, AI may find it difficult to understand human subtleties and context.

Legal and Ethical Issues: Adherence to rules and ethical considerations, particularly those pertaining to monitoring and justice, is crucial.

Dependency on Data: Artificial intelligence depends on high-quality data, and bad data might lead to bad choices.

Employee Resistance: Adoption of AI may be hampered by fear or skepticism.

Ethical Surveillance: Employing AI for surveillance may give rise to trust and privacy issues. For AI to fully realize its promise in HR, organizations need to solve these issues.

CONCLUSION

In conclusion, AI has enormous potential to change HR operations, but its effective integration necessitates a deliberate, calculated strategy. HR directors need to put an emphasis on implementing ethical AI in a transparent and equitable manner, avoiding problems like bias and data privacy concerns. Prior to implementing AI across other departments, HR teams should begin with simple, scalable AI initiatives, such as automating repetitive processes like applicant screening or employee feedback analysis. This will help them gain trust in the technology and improve it over time. AI is only as good as the data it analyzes; thus, making sure the data is of high quality is essential. Decisions will be more equal and accurate if there is a significant investment made in solid data management and varied, impartial datasets. To build confidence between HR professionals and workers, transparency must be a top priority throughout the process, with explicit explanations provided for AI-driven results. Upskilling HR staff is a crucial component in adopting AI. HR workers will be able to analyses AI insights and make data-driven choices with the help of data analytics and AI literacy abilities. In addition, HR executives who get change management training will be more equipped to lead their staff through the organizational changes that AI adoption will bring about, allaying fears of job displacement and promoting teamwork.

REFERENCES

Avolio, B.J., Sosik, J.J., Berson, Y. (2013). Leadership models, methods, and applications: Progress and remaining blind spots. In: Day, D.V., (Ed.), *The Oxford handbook of leadership and organizations* Oxford University Press.

Bughin, J., Hazan, E., Lund, S., Dahlström, P., Wiesinger, A., Subramaniam, A. (2018). *Skill shift: Automation and the future of the workforce.* McKinsey Global Institute.

Bessé, J.E. (2019). *AI and jobs: The role of demand.* National Bureau of Economic Research. [http://dx.doi.org/10.3386/w24235]

Binns, R. (2018). Fairness in machine learning: Lessons from political philosophy. *Proceedings of the 2018 Conference on Fairness, Accountability, and Transparency,* 149-159.

Bersin, J. (2018). *AI in HR: A practical guide for HR professionals.* Josh Bersin Academy.

Van Esch, P., Black, J.S. (2019). Factors that influence new generation candidates to engage with and complete digital, AI-enabled job application processes. *Bus. Horiz, 62*(6), 729-739. [http://dx.doi.org/10.1016/j.bushor.2019.07.004]

Cappelli, P. (2019). Your approach to hiring is all wrong. *Harv. Bus. Rev, 97*(3), 48-57.

Cowgill, B., Dell'Acqua, F., Deng, S. (2020). Biased programmers? Or biased data? A field experiment in operationalizing AI ethics. *SIGKDD Explor, 22*(1), 12-23.

Davenport, T.H., Ronanki, R. (2018). Artificial intelligence for the real world. *Harv. Bus. Rev, 96*(1), 108-116.

Guenole, N., Ferrar, J., Feinzig, S. (2017). *The business case for AI in HR: How to automate and personalize learning and development.* IBM Smarter Workforce Institute.

Manyika, J., Chui, M., Miremadi, M., Bughin, J., George, K., Willmott, P., Dewhurst, M. (2017). *A future that works: Automation, employment, and productivity.* McKinsey Global Institute.

Moore, P.V. (2019). Surveillance and privacy in the workplace: Challenges for law, technology, and industrial relations. *European Labour Law Journal, 10*(2), 185-204.

Pulakos, E.D., O'Leary, R.S. (2011). Why is performance management broken? *Ind. Organ. Psychol, 4*(2), 146-164. [http://dx.doi.org/10.1111/j.1754-9434.2011.01315.x]

Raghavan, M., Barocas, S., Kleinberg, J., Levy, K. (2020). Mitigating bias in algorithmic hiring: Evaluating claims and practices. *Proceedings of the 2020 Conference on Fairness, Accountability, and Transparency,* 469-481.

[http://dx.doi.org/10.1145/3351095.3372828]

Tambe, P., Cappelli, P., Yakubovich, V. (2019). Artificial intelligence in human resources management: Challenges and a path forward. *Calif. Manage. Rev, 61*(4), 15-42. [http://dx.doi.org/10.1177/0008125619867910]

Westerman, G., Bonnet, D., McAfee, A. (2014). *Leading digital: Turning technology into business transformation.* Harvard Business Review Press.

Wright, P.M., Ulrich, D. (2017). The future of HR and the need for change. *Hum. Resour. Manage, 56*(2), 409-414.

SUBJECT INDEX

A

B

C

Rishikaysh Kaakandikar, Keshav Kaushik, Supriya Pathakm Madhavi Shamkuwar & Vivek Mishra (Eds.)

D

E

F

G

H

I

J

L

M

N

O

P

Q

R

S

www.ingramcontent.com/pod-product-compliance
Lightning Source LLC
LaVergne TN
LVHW070948210926
848482LV00002B/139